Mortuary Monuments and Burial Grounds of the Historic Period

MANUALS IN ARCHAEOLOGICAL METHOD, THEORY AND TECHNIQUE

Series Editor:
Charles E. Orser, Jr., *Illinois State University, Normal, Illinois*
Michael, B. Schiffer, *University of Arizona, Tucson, Arizona*

ARCHAEOLOGICAL SURVEY
E. B. Banning

LITHIC ANALYSIS
George Odell

MORTUARY MONUMENTS AND BURIAL GROUNDS OF THE HISTORIC PERIOD
Harold Mytum

A Continuation Order Plan is available for this series. A continuation order will bring delivery of each new volume immediately upon publication. Volumes are billed only upon actual shipment. For further information please contact the publisher.

Mortuary Monuments and Burial Grounds of the Historic Period

Harold Mytum

University of York
York, United Kingdom

KLUWER ACADEMIC / PLENUM PUBLISHERS
New York, Boston, Dordrecht, London, Moscow

Library of Congress Cataloging-in-Publication Data

Mytum, H. C.
Mortuary monuments and burial grounds of the historic period / by Harold Mytum.
p. cm. – (Manuals in archaeological method, theory, and technique)
Includes bibliographical references and index.
ISBN 0-306-48075-1 – ISBN 0-306-48076-X (pbk.)
1. Sepulchral monuments. 2. Cemeteries. 3. War memorials. 4. Funeral rites and ceremonies. I. Title. II. Series.

CC77.B8M96 2004   2003061895

ISBN HB. 0-306-48075-1
PB. 0-306-48076-X

233 Spring Street, New York, New York 10013

http://www.wkap.nl/

10 9 8 7 6 5 4 3 2 1

A C.I.P. record for this book is available from the Library of Congress.

Printed in the United States of America

To Michael Farr, Bill Ford, and Jocelyn Morris, who encouraged me in my archaeology whilst at school and university, and who gave me my first projects in Warwick, from which my interests in historical archaeology came.

Preface

Burial grounds strike an immediate chord with all who visit them. They are landscapes full of pathos and cultural associations which many find attractive, though a minority feel are too morbid to deserve detailed attention. This book is designed to offer a framework for studying historic burial ground monuments, and contains a certain amount of information regarding below-ground archaeology, as some projects will involve the study of both. Moreover, from a research perspective above and below ground archaeology together can throw considerable light on the process of dying, body disposal and commemoration that formed a continuum for those involved. However, the more easily accessible graveyard memorials can be used to study many aspects of past culture beyond that directly associated with death, and they are the focus of the book.

Most people who become interested in historic graveyard memorials come to the subject via the material itself. Only rarely does a research question get posed, and graveyard data seem like an appropriate arena for investigation. This has the advantage that many researchers have some ideas about the data available, but the disadvantage that they may not have clear questions to ask of it, nor how to set their discoveries in a wider intellectual context. The purpose of this book is therefore to show those with particular questions that, in some cases at least, graveyard data may be a valuable research area, but also those with an interest in burial grounds and their monuments what sorts of topics can be studied through their data, and how their results can be compared and contrasted with those from elsewhere. Through a series of chronological chapters (2–4) the general sequence of development of mortuary behavior is outlined, then the material is considered under a series of thematic chapters (5–8). Here the curent state of knowledge and the research questions already posed in at least some regions and periods are outlined. Suggested strategies for developing these aspects, and also new areas are proposed. Thus, the chapters provide a stimulus to directed research and references to relevant literature to help situate further work within the existing knowledge base.

The book could have provided a step-by-step method of recording graveyard memorials, but as the data required for different research questions is so diverse, and the range of forms, types of decoration and cultural context so variable over both time and space, this would have created either an extremely large and unwieldy volume, or one which was problematic and confusing for most users. Instead, emphasis has been placed on the background and context of memorialization, and on the general factors that need to be considered in the light of local circumstances and research aims. Many of the publications give little information on how fieldwork was conducted, but some of the practical issues are discussed here, and examples of recording forms are provided in the Appendix. Chapter 9 does give some practical advice regarding graveyard recording, including some of the specific elements that make historic burial excavation unique.

Graveyard memorials form a rich seam of archaeological evidence, but publication is often in a local format. I have tried to offer an international range of examples to show general trends and the effects of local cultural traits. We are all familiar with our own areas and regions, and imagine others must be the same. Despite some national and indeed global features, many aspects have a clear local manifestation. The situating of graveyard studies within the local is both valuable and stimulating, but it can mean that valuable studies of interest to others are not discovered. I have tried to follow up references and search for appropriate examples, but my inevitably restricted experience may be visible in places, and some books and journals I have not been able to trace from Britain. Some aspects which I consider underdeveloped may have been examined and published in journals not easily available to me, and as work proceeds this will undoubtedly be the case, so any information on other published material would be welcome. The bibliography contains many recent references and some of the classic ones, but I recommend detailed study of Bell's *Vestiges of Mortality & Remembrance. A Bibliography on the Historical Archaeology of Cemeteries* (1994), and subscription to the Association for Gravestone Studies. Some may find my *Recording and Analysing Graveyards* (2000) of value, though written primarily for a British and Irish context.

For too long graveyard studies have been seen as an eccentric by way rather than a specialist subject area with as much to contribute as the study of ceramics, building types or faunal remains. As a category of material evidence with variety of form, decoration and text, and well contextualized spatially and temporally, it should be suitable for many forms of analysis. Excavated mortuary evidence has tended to be seen as a source for biological data, and the cultural dimension has often been underdeveloped. The linking between above and below ground data has rarely been achieved, and the integration of graveyard data within settlement and landscape archaeology has also been likewise rarely attempted. Whilst much has been undertaken, there is far more to do, and it can be achieved on a range of scales appropriate to the resources of individuals and teams. Moreover, as so much of graveyard archaeology is non-destructive, it carries fewer ethical constraints and can be undertaken by those at many stages of their careers. It can be an archaeological activity well placed to be integrated within the community, and an arena where local populations can feel a sense of pride and association with material culture from the past.

I have been involved in graveyard recording for many years, and in that time many people have encouraged my interest, and helped me with my researches. Philip Rahtz was an important early influence, and I thank him for his continued support and the use of some images reproduced in this book, listed below. My wife Caroline was a great help with my surveys in Wales and Gibraltar, where Chris and Christine Webster also gave valuable assistance. More recent surveys in Ireland and Yorkshire have benefitted from the help of Steve Rowlands and Carol Simmonds, and particularly the enthusiasm and dedication of Robert Evans, who has also kindly read through and commented on the text. Most of all, however, graveyard recording is a team effort, involving many people, and all the undergraduate and graduate students from the University of York, international students on the University of York Castell Henllys Field School, and Earthwatch volunteers need to be thanked for their steady application in the field and in data entry and processing. My work on the York Diocesan Advisory Committee has allowed me many opportunities to consider the implications of ecclesiastical development and conservation in relation to below and above ground archaeology, and to see the competing needs of various interest groups in relation to the heritage. Illustration credits: all photographs are by the author except for those by Philip Rahtz (figures 19, left, 22, 29, 42, 45, 46) Joshua de Giorgio (figure 14, left) Catherine Marlow (figure 7) and Department of Archaeology, University of York (figures 4, 5, 40, 53); I thank them for the gift of their images, or permission to publish them here.

This book aspires to create a more diverse use of graveyard monuments, and allow scholars to place their own findings in a wider context. It therefore contains a large number of references, and examples from many places. It is suitable for use by students, by CRM professionals, and those involved in education at a number of levels. Though written from an archaeological perspective, it should have a value to cultural historians, folklorists, geographers, anthropologists, sociologists and art and architectural historians.

I look forward to the developing research agendas of the years ahead, and hope that those who read this book are stimulated to use existing data in new ways, and collect, analyze and publish the results of their own field work. I would be grateful for bibliographic references and where possible copies of published papers so that any further edition of this book can incorporate relevant new work.

Harold Mytum
The King's Manor
York, United Kingdom

Contents

List of Figures

List of Tables

Chapter 1

Introduction

The study of mortuary evidence has a long pedigree in archaeology, and many of the methods and theories applied to historic burial grounds and their monuments can be similar to those of earlier periods. There are, however, many advantages for those working in this particular field within a relatively richly documented context. Death and burial were important events, which had religious, legal, social and cultural consequences, and as a result they often required various forms of documentation. Thus wills, probate inventories and burial registers all are directly associated with death, and are the types of official documents which survive relatively frequently. Moreover, deaths and funerals were occasions which could be recorded in less formal contexts such as newspaper reports, diaries and letters. The historical archaeologist is therefore able to approach the material culture of death with questions informed by this documentation. Archaeologists should not to be constrained by written sources, however, but be able to explore aspects of the past not easily defined by those working in earlier periods.

The very presence of extensive documentation, combined with the concerns with examining relatively recently interred human remains, has hindered the development of some aspects of analysis, and perhaps rightly so. However, an increasing range of studies can now be undertaken, often through development schemes on excavated material. Moreover, the possibilities for studying above-ground memorials and mortuary landscapes is now being better appreciated. This does need to be placed in some historiographical perspective so that the achievements to date, and the potential to come, can both be appreciated.

No large-scale study of the development of historical mortuary analysis has been undertaken, and this is not the place to remedy this omission. Rather, the following headings seek to expand and develop the themes explored by Bell (1994) and Tarlow (1999b) in their historiographical and theoretical reviews, and the citations should be seen as examples. Even from antiquarian times there have been numerous references to material of relevance to this topic, and in recent decades large amounts of material has been produced on the subject. Bell (1994) covers a great deal, though this is very partial outside North America, and the annual review of publications in *Markers*, beginning with volume 12 (1995), has gradually become more global in its coverage. Other useful bibliographies include

Thompson et al. (1993) for the Southern United States, and Thompson (1997) for African American and African Caribbean evidence.

1. A BRIEF HISTORY OF HISTORICAL MORTUARY ARCHAEOLOGY

1.1 Antiquarian Interests

Although human remains of the last few centuries have been disturbed by continued burial, building at churches, and redevelopment of old burial sites, antiquarian interest was limited because the remains were so recent. There was some fascination during the 19th century in Britain and Europe with the burials of the elite, however, and so bishops and royalty were the subject of exhumation and analysis (Rodwell, 1981). Other vaults were also visited and recorded, some in detail (Bigger, 1898). The information was largely of interest to family historians and antiquarians researching particular localities, and little attempt was made at comparative studies. In this regard historical remains were not treated in the same way as prehistoric, Classical or early medieval discoveries.

The inscriptions and heraldry on memorials was considered worthy of record by antiquarians who saw the lineages of the aristocracy as of vital importance. By the 19th century, numerous collections of inscriptions were being made, and many were published. Some were linked to particular places; others concentrated on the messages conveyed in the epitaphs. The interest in old memorials, inside and outside churches, was widespread in Britain, Europe and the east coast of North America by the later 19th century. Some attempts were made to prevent destruction, but many monuments were removed during this period without record. In North America, the memorials provided a tangible link with early phases of settlement, and so they were seen as more important than in Britain and Europe. Thus, for example, in the last decade of the 19th century the Daughters of the American Revolution attempted to conserve and restore Hartford's Ancient Burying Ground, and a large amount of money was raised and expended on this project (Hosley and Holcombe, 1994).

1.2 The Origins and Development of Contemporary Research

It has been during the 20th century that many aspects of research have begun on historic burial grounds. Initial studies have come out of individual curiosity, but have subsequently developed into wider areasof research undertaken by many people. Methodologies transferred from other aspects of archaeology have been developed for the particular types of evidence available in this period, and scholars are now applying a wide range of theoretical perspectives to the material.

1.2.1 Memorials and Carving Traditions

The first half of the 20th century saw little change in the attitude to historic burial markers, and indeed the extent of recording decreased. There were, however, a few pioneers of a more detailed and analytical approach, greatly aided by the now widespread availability of relatively portable cameras, which allowed photography of memorials; those relating to New England are noted by Buckeye (1977). The seminal work on gravestones is that of Harriette Forbes (1927), in which many of the themes were laid out that have been subsequently developed. The role of particular carvers, regional traditions, and the symbolic associations of motifs were all discussed and illustrated. In Britain, research in Scotland also began to identify design features but was still largely in a descriptive site catalogue format (Christison, 1902; 1905; Reid, 1909; 1915). A similar pattern was followed in Ireland, where in Ulster armorial designs were recorded (Bigger and Hughes, 1900; 1901; 1902; 1903).

The interest in carvers and regional styles of external memorial began to grow in the middle of the 20th century, mainly in New England through Caulfield (collected papers republished in 1991), in Scotland (Graham, 1958), and in England with some studies in the Midlands (Barley 1948; Herbert 1944). From the 1960s onward the rate of interest grew rapidly, with seminal studies such as Ludwig (1966) in New England and Burgess (1963) in England.

1.2.2 Scientific Excavation of Historic Burials

In contrast to the increased interest in memorials, attention to excavated remains was still limited. Many excavation projects in Britain and Europe removed interments of the historic period as rapidly as possible to reach what were viewed as the culturally more important medieval and earlier deposits. Where excavation and recording was more carefully undertaken, the results were not published. In America, some excavation and publication took place, but usually because there was a perceived high cultural value as such burials were linked to important phases of the country's history. Thus, Cotter excavated and published the results of excavations at Jamestown (Cotter, 1958), and Noel Hume's discoveries of burials in the context of settlement excavation could be linked to early colonial history (Noel Hume, 1992).

From the 1970s, as the role and scope of historical archaeology grew in North America, and as its place with cultural resource management became better established, excavation of human remains within mitigation strategies became more frequent. Whilst most material has remained in the gray literature (see Bell, 1994 for examples), large numbers of skeletal analyses have been carried out and many published in specialist journals. Some cultural studies have also taken place. A relatively early substantial published example was at the Uxbridge Almshouse burial ground (Elia and Wesolowsky, 1989) which, through various subsequent articles, has revealed the potential of even one low status burial ground (Bell, 1990; Elia, 1992). The advantage of documented sources such as trade catalogues has allowed the identification and classification of coffins and fittings from some sites, allowing relatively rapid movement beyond the traditional archaeological concerns over classification and dating.

In Britain, the Spitalfields project highlighted the potential for historic burial archaeology (Molleson and Cox, 1993; Reeve and Adams, 1993), and investigations linked to development have followed at a number of sites, both in and around churches and at cemeteries. Nevertheless, detailed analysis has been limited, and not with the intensity or level of interpretation reached in North America. An exception would be with regard to the recording of burial vaults, which has continued at a greater level than in North America (Cox, 1998a; Litten, 1999; Mytum, 1988; Redknap, 1984), though even here description has dominated the literature.

1.2.3 Maturing Memorial Research

From the 1970s, the art historical and archaeological potential of gravestone monuments began to be appreciated. The Dublin Seminar for New England Folklife Annual Proceedings (Benes, 1976) provided a forum where many approaches were aired, and in due course the Association for Gravestone Studies was established, with its Newsletter, journal *Markers*, and annual conferences. Another important conference venue for North American research is the annual theme on memorials at the American Anthropological Association. Major monographs based on localised studies appeared at this time (Benes, 1977; Tashjian and Tashjian, 1974), a tradition that is continuing (Luti, 2002). Some wider regional studies have also been produced (Jordan, 1982; Little, 1998). Some specific cemeteries have also received treatment (Linden-Ward 1989a), and many volumes have been produced with a large photographic element (Brown 1994; Florence, 1997; Sledge, 2002).

The search for regional traditions has continued in England (Lees, 1993) and Scotland (Willsher and Hunter, 1978), though with no substantive detailed studies. There has been surprisingly little interest in defining the work of specific carvers (Turner, 1985). This has been attempted in Ireland, however, (Grogan, 1998), building on the work of Longfield (1943), but again no monograph scale products have emerged.

In Britain, the archaeological recording of graveyard monuments has been a significant concern, beginning with the rescue movement which highlighted the destruction of the archaeological resource (Jones, 1976). A small number of site-specific descriptive studies resulted from this initiative (Rahtz and Watts, 1983; Shoesmith, 1980), but many were not able to draw wider conclusions because of the scale of the work. The largest and most comprehensive regional study has been undertaken in Pembrokeshire, Wales (Mytum, 1994; 2002e), though Cannon

(1989) in Cambridgeshire, England and Tarlow (1999b) in Orkney, Scotland, also utilized substantial samples. These programs have allowed the investigation of broader questions of identity, variability, and emotion. Methodological concerns regarding recording and analysis have also developed (Mytum, 2000; 2002b). No archaeological recording programs have been undertaken in Europe, apart from one on the Protestant cemetery in Rome (Rahtz, 1987) and Gibraltar (Mytum, 1993).

2. THEORETICAL APPROACHES

A large amount of research into memorials and burials has been carried out within a generalized art historical or historical context, with no explicit theoretical framework. On examining the aims, methods, and conclusions drawn in such work it is usually clear that the research was based around what archaeologists would term a culture-historical approach. This largely inductive and descriptive endeavor often contains elements of functionalist interpretation, and contextualist explanations also are widespread. Some academic studies, however, are well grounded in theoretical traditions from art history or folk life studies, and some of the lines of inquiry so highlighted could be usefully explored further in an archaeological context.

2.1 Culture-history

Until the 1960s the dominant approach to mortuary remains was that of description, with limited interpretation being constructed within a culture-historical paradigm (Childe, 1956). Excavation of human remains is often still also undertaken within a cultural historical approach, with the aim of identifying through grave goods and aspects of the burial tradition the cultural affiliations of the deceased. Given the relatively well documented affiliations and population movements of recent centuries, historic burials have attracted less attention by those interested in this approach.

This style of research has remained extremely popular for those interested in funerary monuments, and in most if not all parts of the world is still the dominant framework within which gravestone recording and analysis takes place. Much of the work is undertaken by local amateur groups or individuals who often have very great expertise and background knowledge of the locale and the history of that place. Whilst the analysis may therefore lack wider comparative approaches, there can be a substantial body of data already collected, checked and ordered which can form the basis for further study.

Antiquarian and genealogical concerns were foremost for early gravestone recorders, who concentrated on the memorials of the elite and were more assiduous in providing details of inscriptions than of form or decoration. Such studies may now be important where erosion and destruction has removed evidence, but have otherwise largely been superseded by more thorough recording and analysis. Genealogists are still very active in many areas, recording names and whole inscriptions. These are largely deposited in local or regional repositories, though some are used in commercial genealogical services, and this has become part of the heritage industry in some countries such as Ireland (Grenham, 1992). Many archives and search facilities are becoming available on the World Wide Web; indeed genealogy is now the second-most popular search subject on the web after sex, a 21st-century commentary on death and sex, supposedly the two great taboos of the 20th century! Whilst lacking the details of the material culture, such collections of inscriptions can point researchers to sites with assemblages of the appropriate period for study.

Some case studies are published which include considerable description or discussion of material aspects of memorials as well as inscriptions. For example, the publications on Paramatta cemeteries, Australia, list all inscriptions but also details about the form, decoration and material of each monument (Dunn, 1998; 1991). Detailed analytical studies can use both text and motif (Tashjian and Tashjian, 1974), though many works of this kind are designed for a less academic audience (Lindlay, 1965; Millar, 1994; Wilsher, 1996). This interest in the combination of named individuals and design features of memorials is often the dominant theme in interpretation as discussed in Chapter 10.

Most academic studies that have concentrated on memorials as material culture have applied the principles of art history. In Europe, internal memorials have attracted the most attention (Llewellyn, 2000; Kemp, 1980; Penny, 1977; Potterton, 1975; Whinney 1988). These still form a major theme in current work, exemplified by most of the papers in the journal *Church Monuments* which published its first annual volume in 1996, and includes British, some European, and a few Colonial studies (Coutu, 1997; Hammond, 1999).

With regard to external memorials, there was at first a tendency to concentrate on folk art and early material, both in America (Caulfield, 1991; Forbes, 1927; Ludwig, 1966), and in Britain (Brears, 1981; Burgess, 1963). This emphasis has continued, notably in New England (Chase and Gabel, 1997), Britain (Chater, 1976; 1977; Elliott 1978; Willsher and Hunter, 1978), and in Ireland (McCormick, 1976; 1983). Many New England analyses in particular have been concerned to identify particular craftsmen (see Chapter 5), whilst others have used gravestones to explore symbol and meaning (see Symbolic approaches below and Chapters 7 and 8).

Gradually more studies of later 18th- and 19th-century memorials have been produced, as architectural historians have appreciated some of the most elaborate structures (Colvin, 1991; Curl, 1980). There has been a widening of interest to include the smaller but particularly evocative memorials, and even 20th-century monuments have been accepted within such studies (Brown, 1994; Florence, 1997;

Keister and Cronin, 1997; McDowell and Meyer, 1994). There has been a particular focus on war memorials, with recent research partly caused by the increased interest derived from the 50th anniversary of the end of World War II (see Chapter 4).

Culture historical approaches have tended not to be explicitly theoretical, and have often concentrated on description with just occasional ad hoc explanations. For those with such theoretical inclinations there is a rich potential to examine and analyze the cultural choices demonstrable in the material culture record. Moreover, as work is conducted more widely around the globe, major zones with different mortuary traditions in the historic period are becoming more obvious. As some of the subjects most closely linked with culture historical study of memorials such as architectural and art history are themselves becoming more theoretical, it is possible that this approach will be transformed into non-functional paradigms such as structuralism and contextualism.

2.2 Functionalist Approaches

As New Archaeology widened the perspectives of archaeology, so attitudes to mortuary evidence changed considerably, and prehistorians used old and new material to present social and economic interpretations previously considered inaccessible. There was an early impact on gravestone studies through the work of Deetz and Dethlefsen, both together and individually (Deetz, 1977, 64–90; Deetz and Dethlefsen, 1965; 1967; 1971; Dethlefsen, 1969; Dethlefsen and Deetz, 1966). This did not lead to a great explosion in further published research in this genre, however, though see Gorman and DiBlasi (1976). Dethlefsen continued to study mortuary evidence (1981; 1992), and Deetz has emphasized a structuralist slant to the shift in motifs over time, and linked this to parallel changes in culture (Deetz, 1977; 1983), discussed further below (see 2.3).

The limited application of functionalist explanation within historic mortuary studies has been well reviewed by Bell (1994, 12–17). Where a wide range of sources is available, simplistic assumptions about the material correlates of ranking can be seen to be problematic, though some attempts at incorporating historic graveyard data in cross-cultural comparisons have been attempted (Cannon, 1989). Social status was indeed a major factor in memorial selection, but more subtle and context-dependent approaches are normally required to identify this factor amongst many.

Attempts at using gravestone data as a basis for demographic study assume that there is a constant relationship between population and memorialization (Dethlefsen and Deetz, 1967) which is clearly not the case. Tarlow (1998; 1999b) notes a 'gravestone boom' in the late 18th century which is not related to population growth. Where documentary sources such as burial registers survive and be compared with the memorial data, then there are clear trends over time (Mytum 2002e). Research could further examine the social and economic complexion of those commemorated, compared with the total population. It is highly likely that the nature and proportion of commemoration varies from region to region, and indeed from one context of burial to another, so that small town colonial burial grounds will have a different commemorative pattern to large urban cemeteries.

The relatively short time spans and detailed context of many gravestone studies usually make alternative approaches more appropriate. Indeed, the very presence of named individuals draws scholars into more personalized studies, perhaps to the detriment of ignoring some larger scale trends, which a processualist approach could provide. These large scale normative trends could then inform more focused studies from any theoretical perspective. Functionalist studies still have much to offer, at the level of production and distribution, and in terms of defining large scale social, ethnic and cultural variation. Where possible these should link down to smaller scale studies, which may allow other non-functional aspects to play their part within the wider trends. Gravestone studies are particularly suitable for multi-level analysis and interpretation.

2.3 Structuralist Approaches

The role of deeply rooted, unconscious sets of rules regarding cultural behavior have been examined by structuralist anthropologists and archaeologists, and death has been a popular subject of analysis. Structuralists are particularly interested in liminal moments and situations, and death provides one such opportunity to examine transition from one state to another, and how this natural event is culturally constructed if not controlled. Thus, many anthropological studies have taken this approach (Humphreys and King, 1981; Huntington and Metcalf, 1979) and historians have also begun to appreciate its potential (Ó Crualaoich, 1998).

A few influential studies taking this stance have been carried out on memorials. Most well known to archaeologists is the work of Deetz, linking changes in belief brought about by the Enlightenment to material changes which included gravestones (Deetz, 1977; 1983). Also significant is that of Llewellyn (1991) who has considered a wide range of elite material culture associated with death, and applied an explicit structuralist model to English internal memorials (Llewellyn, 2000).

2.4 Marxist Approaches

Traditional Marxist analyses have been popular in historical archaeology, in part because of shared concerns over the rise and functioning of capitalism (Orser 1996). Graveyard material has been interpreted within this paradigm most notably in the New York area on material produced within a developed capitalist system (Clark, 1987; McGuire, 1988; 1991; Wurst, 1991). Another much-quoted study of Cambridge, England, examined aspects of funerary behavior but with limited attention to memorialization (Parker Pearson 1982). War memorials could be a source of tensions between classes in their design and siting, and could act as focal points in later protests (Mansfield, 1995).

Marxist approaches have tended to concentrate on the false consciousness which can be identified through memorial designs and texts. They have revealed the contradictions between the lives and experiences of those commemorated, and the idealized roles portrayed on the stones. Other aspects of research that would be of considerable interest would be the individual cases and general prevalence of resistance to this form of expected commemoration. In contemporary cemetery management there is plenty of evidence for conflict between grieving families and the authorities, and it is likely that this could be explored for at least the 20th and parts of the 19th century. The false consciousness within elite memorials has also not been greatly explored, yet this is an essential part of the whole class structure, which needs examination.

Another theme that deserves much more attention is one of the production and distribution of memorials. The effects of capitalism on changing the scale and nature of production, and the effects of industrialization on the place of the craftsman carver, have been noted (Mytum, 2003c; Petke, 1993) but not explored in this paradigm. As memorials continue to have an element of personal hand-crafted production up to the present day, and involve the interaction of producer and client, they offer an insight into this dynamic in a world dominated by mass production and consumer choice only as purchaser.

2.5 Symbolic Studies

Long before post-processual archaeologists had begun to examine the complex relationships between material culture and meaning, the more analytical culture-historical scholars were deeply immersed in a contextualized reading of gravestone art. Whilst not phrased in the jargon of later paradigms, they voice similar concerns with the active role of material culture, the carrying of multiple meanings, and the manipulation of symbol. The Puritan world view and its effect upon gravestone designs, and the ways in which rival interpretations could be played out through motifs was a concern of several major studies (Benes, 1977; Tashjian, 1978; Tashjian and Tashjian, 1974; Watters, 1978; 1981). Likewise, the various strands of churchmanship within the 19th-century Anglican church and other denominations led to debates over the use of Gothic, Classical or other revival styles (Curl, 1980; McDowell and Meyer, 1994). Symbolism relating to social structures, (Chapter 6), identity (Chapter 7), and attitudes to death and resurrection (Chapter 8) have been some of the major themes in historic mortuary archaeology.

Historical archaeologists have long been carrying out contextual archaeology without using that term, but prehistorians such as Hodder (1986) have explicitly developed this theoretical approach. This has highlighted the active role of material culture which can be acknowledged in other paradigms but which is given greatest prominence here, with an emphasis on the range of meanings that items may have to different individuals and groups. This has been an important part of mortuary archaeology, both in the study of graves and memorials, but particularly in the latter for the historic period. Some studies acknowledge the newly defined contextual link (Burley, 1995; Tarlow, 1999a; b), but many others continue to apply it in ways appropriate to the existing literature and research framework. Finch (2000) has provided an archaeological approach to internal monuments, a useful contrast with the art historical perspective, and one which offers a contextualized analysis of changes in commemoration over several centuries.

The symbolic value of designed cemetery landscapes has been a popular subject with architectural and landscape historians examining cemeteries in Britain and North America (Brooks, 1989; Linden-Ward, 1989a; b; Sloane, 1991). The conflict between these western perspectives and those of the Chinese in Singapore highlights how such western attitudes have been applied widely over the globe in colonial contexts, and the reactions of those with other beliefs (Yeoh, 1991; Yeoh and Hui, 1995). The theme of the "beautification of death" is a factor in cemetery design, and this has been developed for funerary ritual as exemplified through coffin design and fittings (Bell, 1990; McKillop, 1995). This concept often lies implicit behind many of the studies of graveyard art, but is rarely developed in any detail (Snyder, 1992). Perhaps more explicit contextualist theory would here highlight ways in which these relationships and meanings could be explored. The possibilities can be seen, for example, in the contextualized study of the relationships between consumer and producer in a cemetery explored by Buckham (1999).

War memorials have also been examined to reveal subtle sets of meanings (Inglis, 1998). The Tomb of the Unknown Soldier at Arlington National Cemetery has also been considered in symbolic terms (Ingersoll and Nickell, 1987), exploring the tensions between society and self within modern society, a theme also examined through British war graves (Tarlow, 1997; 1999b). American war cemeteries abroad were designed to represent that country on foreign soil, with intriguing choices of style and symbol (Robin, 1995).

2.6 Other Recently Developed Approaches

The emotion of the bereaved has been highlighted by Tarlow as a major factor in the development of external memorialization (1998; 1999a; b). She sees increased interest in individual love and affection being manifested in a changed attitude to the death of close relatives in the later 18th and 19th centuries, explored at least in part through empathy.

Hermeneutic theory has been applied only rarely to graveyard monuments, but Garman's study (1994) of memorials at Newport, Rhode Island is one such study. Phenomenology within cemeteries and burial grounds could be a very productive line of inquiry. This could also be applied to an exploration of the funeral experience, with consideration of different elements through time starting with sitting with the corpse, then processing to church, the internal service, and thence to the burial site and the final interment, followed by funeral meals and continuing mourning and grave visiting. This involves private experiences and public events in a variety of locations, the physical and social environment of each creating experiences for the bereaved. Feminist perspectives have been offered on Australian war memorials (Inglis, 1989) and the Punchbowl war cemetery, Honolulu (Ferguson and Turnbull, 1996).

3. SUMMARY

The overall position regarding the archaeological use of historic mortuary evidence is patchy, though there are many examples of excellent research on particular sites or assemblages. Given the potential of the material evidence, often containing text, symbolism, and many other material attributes linked to well documented individuals and circumstances, data from historic burial grounds has not yet been accepted in a central place within the sub-discipline. Settlement and domestic artifact evidence is overwhelmingly chosen to research wider questions to which mortuary evidence could provide valuable input.

Chapter 2

Folk Traditions and High Culture

Funerary and Commemorative Practice to the Early 18th Century

The earliest part of the historic period is the one which has proved most problematic to study from an archaeological perspective. Relatively few burial grounds from this period have been excavated, with interments of this date in Britain and Europe often disturbed by later burials, and being relatively rare in colonial contexts. Memorials are much more accessible for study, but were erected relatively rarely and only by a restricted segment of the population.

Many British and European memorials are internal, with relatively few external stones surviving. In North America, external stone memorials dominate the surviving record, but even these need to be placed within a wider context of death, burial and commemoration. It is thus necessary to begin with attitudes to death, followed by the form, function and symbolism of the funeral, and finally to consider commemorative practices, so that the elements of surviving material culture can be seen as active in different parts of this set of cultural meanings and activities.

Religion was a major structuring principle for beliefs and practices regarding death during this period. Whilst religious figures were experienced in providing death rituals, and the College of Arms were expert in elite funerals, most aspects of mortuary culture were provided by non-specialists. Thus at this period material culture helped to construct the popular folk behavior associated with death, burial and commemoration. Even specific products such as coffins were made using items such as nails and handles used in furniture production or lead necessary for plumbers, and most memorials were produced by craftsmen mainly involved in building and architectural carving or furniture manufacture.

1. DEATH

Throughout the 16th, 17th and early 18th century, death rates were relatively high. Whether in crowded urban environments, hostile colonial settings, or in rural landscapes where poverty was rife, life expectancy was no higher than it had probably been in the Middle Ages. Relatively few burials have been excavated for this period, and most demographic studies are based on documentary sources such as burial registers. Social historians have studied these records extensively, and these provide a valuable context for the study of death and burial (Dobson, 1997; Slack, 1985; Wrigley and Schofield, 1981).

1.1 The Good Death

One of the obsessions in the Protestant world during the whole of this phase was to have a good death. To achieve this, there had to be demonstrations of piety, worldly concerns had to be left in good order, and one should be awake at death (Beier, 1989; Gittings, 1999; Houlbrooke, 1998; Llewellyn, 1991). In the Catholic world, absolution by a priest at death was the essential requirement, but other features such as settling one's affairs were similar to those of Protestants (Tait, 2002). In all cases, therefore, preparation was essential, and concerns over death loomed large in the lives of families of all classes. This concern with the good death had little material consequence at this period, and so cannot be explored extensively in archaeological terms, though the paintings and engravings of deathbed scenes are informative regarding the idealized physical context desired for the good death (Lewellyn, 1991).

2. THE FUNERAL

For Protestants, funerals were aspects of memorial, as salvation of the deceased was based on his or her faith, and in some sects on their deeds or the belonging to an 'elect' (Cressy, 1997; Stannard, 1977). In contrast, Catholic doctrine emphasized the continued role of prayer of those still living in aiding the deceased in the process of passing from Purgatory to salvation; this gave funerals a purpose in beginning this process. Funerals in England could be extremely elaborate for the gentry and above who would require a heraldic funeral (Litten, 1991; Wagner, 1967), a tradition also continued in some parts of America. Even amongst the elite there was dissatisfaction with this style of ceremony, and night funerals became popular, beyond the control of the vested interests of the heralds (Litten, 1991). There was little music, and Henry Purcell only reworked an earlier composition for royal funeral of Mary II in 1695; there was still little used more generally until well into the 18th century (Llewellyn, 1991:83). From the 17th century onwards, funeral sermons became more common and, if subsequently printed, acted as a form of memorial (Houlbrooke, 1998:295–330; Tashjian and Tashjian, 1974:37–48).

By the late 17th century, a new occupation of undertakers began to compete with the heralds for high class funerals, and they widened social access to aspects of the funeral repertoire (Fritz 1981; 1995). There was widespread giving of food or small sums of money to those attending funerals (Houlbrooke 1998:264–294);

in Catholic contexts this increased the numbers attending the funeral and so the prayers for the deceased, but it had only a social aspect throughout the Protestant world. Gifts to close family and important guests could include items of mourning clothing, spoons or jewelry, especially rings (Bury, 1986; Tashjian and Tashjian, 1974:27–30).

2.1 Mourning Dress and Funeral Expenditure

Sumptuary laws controlled mourning dress in England, though as with other aspects of funerary conventions, these were increasingly ignored and extravagance became widespread during the early 18th century. By the late 17th century, quantities of a silk called mourning crepe was being made in England by Hugenot weavers, a cheaper version of imported materials only used previously by the aristocracy (Llewellyn, 1991:91). Mourning dress involved the use of black and other dull colors such as dark green or gray (Cunnington and Lucas, 1972; Taylor, 1983). After a period of full mourning, an intermediate state of half-mourning was expected. This involved adapting normal clothing, and using some items such as black and white fans, popular in the early 18th century (Hirschorn, 1973). Throughout this period, and indeed after mourning was over, memorial jewelry could be worn, particularly rings (Bury, 1986). Even in puritan New England, a similar range of funeral expenses are noted in accounts, and here the funeral was normally at sunset. Cakes and alcoholic drinks were dispensed at such occasions (Benes, 1977:33–37; Stannard, 1973). The extravagance of funeral expenditure caused offense in some colonial circles, and in Massachusetts sumptuary laws were first passed in 1721 to restrict the amount of gifts at funerals (Benes, 1977:225).

For the funeral, coffins were often covered with a pall of appropriate quality rented for the purpose. For most people, a bier carried by two people was used, but the more affluent were transported by horse-drawn hearse, decorated according to status, a form of status display from home to church (Llewellyn, 1991). Poorer funerals very much varied according to resources, and many had no coffin but were taken to the graveside in the public coffin provided by the parish which could be reused (Litten, 1991). In colonial frontier situations communal recognition of a death was not possible on isolated farmsteads, but was still achieved in the more densely settled countryside and in towns.

3. BURIAL GROUNDS

Communal burial was the normal practice during this period, with formal enclosed areas being set aside for this purpose (Table 1). Most were under the control of a religious authority, though gradually non-sectarian community burial grounds were founded. Only a few burials were outside the burial enclosures.

Table 1. Attributes of burial grounds.

Name	Date	Location	Layout of graves	Design	Most common monuments	Ownership	Distribution
Churchyard	9th–20th c	Around church	Organic, plots	None, garden	Simple, iconographic, sculpture, plots, none	Religious	UKI, NA, Col, Eur
Frontier graves	17th–20th c	Site of death	None	None	None or simple	None	NA, Col
Homestead graveyard	17th–20th c	Farm field	Irregular rows	None, garden	Simple, iconographic, none	Family	NA, Col
Potter's Field	17th–20th c	City edge	Regular rows	None	None or simple	Public	NA, Col
Town/city cemetery	17th–20th c	City edge	Regular rows, plots	Formal garden	Simple, iconographic, sculpture, none	Public	UKI, NA, Col, Eur
Rural cemetery	1831–1870s	City edge Suburb	Rows, plots	Garden, picturesque	Simple, iconographic, sculpture, mausolea, none	Private co, public	UKI, NA, Col, Eur
Necropolis cemetery	19th–20th c	City edge	Loculi, rows, plots	Built-up, garden	Loculi plaques, simple, sculpture, mausolea	Public	Eur
Chapel burial ground	19th–20th c	Near to church, at distance	Regular rows	Formal, garden	Simple, iconographic, plots	Religious	UKI, Col, Eur
Churchyard, chapel ground extensions	19th–20th c	Near to church, at distance	Regular rows	Formal, garden,	Simple, iconographic, plots	Religious	UKI, Col, Eur
Lawn-park cemetery	1855–1920s	Suburb	Rows, plots	Pastoral, park-like	Simple, sculpture, small monuments	Private, public	UKI, NA, Col
Memorial park	1917-present	Suburb	Rows, plots	Pastoral, suburban	Ledgers, plaques	Private	NA, Col

Key: UKI = England, Wales, Scotland, Ireland, NA = North America, Col = European colonies, Eur = Europe. Developed from Sloane, 1991.

The most frequent were criminals and, in Catholic countries, those who had been excommunicated. Where dissenters of any kind wished to be buried in unconsecrated ground, in contrast, this often led to prosecution (Gittings, 1999). Many colonial regions were also dominated by the established church, whether Protestant or Catholic, though many settlements were established by those of independent sects; they were not necessarily any more accepting of views other than their own, however, and in this phase religious tolerance was rare.

3.1 Location

Location of burial was influenced at this period by religious geography, with one denomination in a controlling position in any one region. In some places, minorities were tolerated, but in most they were not allowed any alternative burial grounds. Depending on cultural traditions, burial grounds were associated with places of worship or were in isolated locations. There is still much potential for necrogeographical studies in this period, placing burial sites within the land use and settlement pattern.

3.1.1 British and European Burial Grounds

In Britain and Europe, most burial grounds were in the form of parish churchyards, which were defined by walls, hedges or banks, and varied greatly in size. There was no significant difference between Catholic and Protestant burial grounds. The form and nature of the burial ground, and its relationship to settlement, depended upon its date of establishment and the complex ecclesiastical and secular history of that area (Bailey, 1987; Morris, 1989). Churchyards normally included the current parish church within their boundaries, known as the curtilage. This was the normal medieval pattern (Hadley, 2001:17–55), but occasionally parish graveyards were not physically associated with the church in use. There were also private chapels belonging to the aristocracy and some institutions such as educational and religious colleges, which had their own associated burial areas. A few nonconformists (notably the Quakers) did establish burial grounds, and Bunhill Fields in London was one example, which began in about 1665 (Meller, 1985:88–93).

Catholic burial could take place at a number of locations, of which the parish graveyard was merely one option. The Counter-Reformation led to the growth of many religious orders such as Franciscans and Jesuits, and these also offered the services of absolution and extreme unction, and burial in their own graveyards. Given that this was a lucrative source of income, and was often supplemented by payments for intercessory prayers to reduce the time spent in Purgatory, there was considerable competition for influence over those in the process of dying (Tait, 2002). This could lead to violence, and even the disinterment of remains and their translation to a different burial location, though by the end of the period the parochial burial structure had managed to reestablish its dominant position.

3.1.2 North American Burial Grounds

In colonial North America, formal burial occurs in three main locations: individual farmsteads, community burial grounds, and in and around parish churches where burial was similar to that in Britain and Europe. Many burials were unmarked, but some of these have been recovered through excavation.

Examples of the farmstead burial grounds have been recovered in the Chesapeake, where burials occur in clusters not far from the buildings. As these are derived from only one or at most a few farms, the number of interments in each burial ground is small (Gibb, 1996), a feature also to be seen in later burial grounds in areas such as New England where memorials can be seen. It is therefore likely that this was a widespread form of burial pattern across the newly colonized areas.

The location of community burial grounds in New England has been studied by Brooke (1988) who notes that 17th-century burial grounds were perhaps as much as a kilometer away from community centers with their meeting houses. The routes of funeral processions in the 17th century traveled from the home of the deceased to the burial ground, where a civil ceremony was performed; there was not an overt religious element. As populations moved out into more dispersed settlement patterns in the late 17th century, however, the burial grounds became more centrally located, and funerals included a religious element including prayers and sermon at the meeting house. Now the burial ground and meeting place were to be adjacent, together with the training field. In new settlements this pattern was established from the beginning, but older settlements could be re-planned. It is noteworthy that this shift in burial location, and its greater prominence and ease of access, also coincided with the appearance of permanent stone markers (Brooke, 1988:467). There was not this shift in Europe, when markers appeared at about the same time, because burial was already around the church and easily accessible. Therefore this would not seem to be the only reason for the development of permanent memorials, though it may have been a factor in the widespread acceptance of this innovation in New England.

Few early church sites have been located and excavated in North America, but the successive churches at Jamestown, Virginia, showed numerous interments within the church as well as in the adjacent graveyard (Crowell and Mackie, 1990:107). Two internal burials were marked with monuments. One was for a knight which had brasses set into the memorial, as was common in Britain at this time (Crowell and Mackie, 1990:123); the other was for a cleric.

3.2 Spatial Arrangement

Intra-site spatial patterning can be elucidated either from extensive excavation producing a grave plan, or by the distribution of memorials. Unfortunately few extensive excavations have been undertaken for this period, and survival of memorials, particularly *in situ*, is rare. Nevertheless, the fragmentary evidence from many sites, combined with documentary sources, does provide some important indications for intra-site patterning.

3.2.1 British and European Burial Patterns

In Britain, burials normally took place within formal burial grounds, and almost all were within churchyards (Rodwell, 1981). The most important would be buried within the church or around its porch; internal burial could be in earth graves or, from the late 16th century, within a family vault for the most affluent (Litten, 1999). Most of the population was buried outside the church, mainly on the southern side, but also to the east and west. The north was almost always avoided, linked to a superstition that this was the Devil's side of the church, though it might be used for those who had suffered 'bad deaths' such as suicide (Gittings, 1999:150). The costs of burial in different areas were often recorded in church records. For example, in 1613 at St. Saviour's, Southwark, London, burial in the church cost 20 shillings, in the churchyard around the church two shillings if in a coffin and 16 pence if not, and in the College burial ground 12 pence if in a coffin and 8 pence if not. Other costs for the gravedigger and bearers would be added. Moreover, those interred within the church could avail themselves of various officers such as the Clerk and Sexton, and could rent a coffin cloth (Brickley et. al., 1999:6–7).

Orientation of burial was almost always east-west, though exact alignment varied and was influenced by nearby physical features such as walls, paths and other burials. There was much inter-cutting of graves, and disarticulated bone was either left in the soil (figure 40), piled up in the corner of newly dug graves, or placed in charnel pits. The intercutting was partly due to lack of space, but also because certain areas within and near to the church were most desired (figure 53). Given the many centuries of burial in these locations, intercutting does not imply ignorance over the location of more recent burials or absence of order within the churchyard, but rather that the older burials could be disturbed to satisfy contemporary demands.

In Catholic Europe the continued medieval practice of storing human bones uncovered during later burial in special ossuaries or charnel houses was widespread (Ariès, 1985; Musgrave, 1997). The arrangement of bones within such ossuaries was, under the influence of the Counter-Reformation and its emphasis on mortality, often greatly altered. Human remains could now be arranged carefully, with skulls placed in rows in elaborate patterns with other bones used to create designs (Colvin, 1991:225). A few medieval ossuaries remained in use in Protestant Britain and even here some structuring of the remains took place.

In the early 18th century, elite burial locations continued in familial church vaults or moved out into the churchyards where sometimes elaborate external memorials could be erected. Some aristocracy in Europe preferred to commission and construct mausolea for family burial on their estates, led in England by the example designed by Hawksmoor at Castle Howard, but occurring also elsewhere in Europe (Colvin, 1991; Curl, 1980). Gardens were also constructed as memorials, usually separate from the burials themselves, and emphasizing a melancholic contemplation of life and death, linked to remembrance of worthy lives. The grand example of this phenomenon was at Stowe, Buckinghamshire, which subsequently inspired further large scale developments on the Continent, such as those at Ermonville in France (Etlin, 1984). These gardens were in turn to have a profound effect on the later development of cemeteries.

3.2.2 North American Burial Patterns

Arrangement of interments within colonial American burial grounds was not uniform. In the case of Chesapeake farmsteads, orientation was not always east-west, and where it was uniform this related to visible features such as buildings or fences. At different phases, there could be some consistency, but favored orientation and layout could vary over time. At Clifts Plantation, what appears to be the tenant's family was in one small group, the others being set at over 10 m away, and most of these were probably of African ancestry (Aufderheide et al., 1981). All interments were parallel and roughly in a line set against a fence. The evidence from Patuxent Point reveals several clusters, the exact number depending on which factors are used to group the burials (Gibb, 1996:196–202). Again, the spatial distinction of tenant families and servants can be suggested, but with more variability in orientation and even some limited intercutting of graves (King and Ubelaker, 1996). Other evidence from College Landing #7 and St. Martin's Hundred Site A also exhibit clusters with varied orientation (Gibb, 1996:230; Noel Hume, 1992).

Burials in and around colonial churches, and in organized burial grounds, conformed to the east-west orientation, and there could be some intercutting in desirable locations, though without the ancestry of burial this was never as severe as in Britain. In Virginia, burial within the church and especially in the chancel reflected high status, as in Britain, and in Bruton church the costs are known. Burial prices were stated in pounds of tobacco, being 4,000 for the chancel, 500 for elsewhere in the church, and just 10 in the churchyard (Crowell and Mackie, 1990:122).

African-Americans may also had their own burial grounds, and these could preserve a range of rites which may have derived from Africa, or were adaptations to the new constrained and exploitative conditions on the plantations and early urban locations. However, most evidence concerning African-American culture comes from later 18th and 19th-century documentation and excavation (Blessingame, 1972; Genovese, 1974). Archaeological evidence for early slave burial is limited, and at Patuxent Point, Maryland, it would seem likely that an African American was interred in the same burial ground as the farm owners (King and Ubelaker, 1996). Although the Newton plantation on Barbados was founded in the later 17th century, probably few of the burials belong to the period up to the early 18th century (Handler and Lange, 1978:132–158). However, it would seem likely that Mound 1 was constructed in the early period of use of the plantation (Handler and Lange, 1978:111), and this may have been some form of foundation burial feature as only one interment was recovered from excavating three-quarters of the mound. In contrast the adjacent Mound 2 was repeatedly used for burial over a long period. Some burials not under mounds produced clay pipes with early 18th-century dates, but most were later. Unfortunately, most plantation documentation gives little or no indication of the location of slave burial areas, though there are hints that interment near the mill yard or the slave accommodation was typical; unfortunately no estate maps mark burial sites (Handler and Lange, 1978:173).

3.3 Burial

Most interments at this period were in simple rectangular dug earth graves, with the body wrapped in a shroud; in England this had to be of wool according to statutes of 1660 and 1678 (Houlbrooke, 1998:341), introduced to increase English wool consumption. These shrouds completely enveloped the body, and were tied head and foot; these were frequently depicted on external memorials, particularly those in New England.

More affluent members of the community were buried in coffins, and these were assembled by local workmen, not by specialists, using hand-made nails. Burials were generally shallower than at a later date, and for that reason and because of frequent overburial, few earth burials survive archaeologically. The accounts of St. Saviour's, Southwark, London in 1613 state that coffined adult burials should be five and a half feet deep, without coffins four and a half feet. Children with coffins were interred four feet deep, and those without only three and a half feet (Brickley et al., 1999:7).

Coffins became more common amongst the English middling classes during the 17th century according to documentary sources (Gittings, 1984:61,114), and by the early 18th century even many of the poor had coffins of their own and the public coffin went out of use (Houlbrooke, 1998:339). Relatively few coffins of this period have been identified in archaeological projects, but studies of British church vaults enables a larger sample to provide a framework, albeit with an emphasis on the more elaborate forms (Table 2). Litten (1985) notes that many of the early coffins were cleared from vaults, and the human remains placed in charnel pits or cisterns within the vault, but some survive and these can be compared with illustrations from the period. The earlier coffins, in use up to about the middle of the 17th century, demonstrate a range of forms, though most were not the hexagonal single-break coffin plan, which became almost ubiquitous later. The anthropomorphic form was only possible because of the malleable nature of lead, but this gave an opportunity for incision of text or addition of lead plaques with details of the deceased (Litten, 1991:92–94). Beyond the basic mummy-like shape, the anthropomorphic coffin could be decorated in various ways, most notably with lifelike three dimensional faces (Litten, 1999:119). The other lead forms all were used in conjunction with a wooden shell. The all-wood gable-lidded coffin form is rare in surviving examples but frequently depicted in contemporary sources and on internal monuments (Litten, 1985:11). One example of the early 17th century was revealed at St. Peter's, Exton, Somerset; made of elm, it had been covered with fabric and decorated with black-headed upholstery nails (Litten, 1991:97). The parish coffin from Easingwold, North Yorkshire, has a slightly gabled lid.

In North America, the excavated evidence for coffins suggests a similar variety of forms to those in Britain, though the numbers excavated are relatively small for this period. Coffins were common, but perhaps the majority of burials were not so encased. It is likely that almost all burials were placed in a shroud, often indicated archaeologically by the presence of shroud pins. Grave cuts varied in length and width, the former often dependent on the age of the deceased.

With coffins came changes to the preparation of the body, with some of the richest families including limited embalming. Dressing the body in specialist funerary clothing also began at this time for those who could afford it; the limitations of only using wool could be avoided by making a payment, and this led to a variety of elite burial garments in other fabrics (Houlbrooke, 1998:341–342).

4. INTERNAL MEMORIALS

In Europe, internal memorials were an important opportunity for display through investment in elaborate art forms, which would be seen by much of the populace on their regular weekly visits to church. Given the protection of the building, there was greater freedom in construction, materials, and use of color than for external memorials. Whilst minor gentry placed monuments in churches under their patronage, major political and ecclesiastical figures were commemorated in cathedrals or nationally significant locations, such as Westminster Abbey in England (Llewellyn, 1996). Many of the aristocracy and royalty built funerary chapels onto existing ecclesiastical buildings or erected new structures for their remains; and these and more modest gentry and mercantile family chapels occur widely across Catholic and Protestant Europe (Colvin, 1991). They provided suitable spaces within which individual and dynastic monuments could be erected.

4.1 Materials

Most memorials were made from stone, with brass insets where stone was not sufficiently fine to carry inscriptions and decorative detail. More elaborate monuments could be partially or completely made from materials transported some distance. In England this was the case with alabaster quarried in the Midlands and used throughout the country though most of the workshops for the finest products using this material were based in London (Llewellyn, 1991). Imported stone of various colors, termed 'marbles', was also used to create polychrome effects, which were further enhanced by the selective use of paint and gilding. In some areas, monuments of similar appearance were created using carved wood and plaster (Kemp, 1980).

4.2 Forms

Internal memorials can be usefully divided into those mounted on the wall, and those set on the floor, and the most elaborate could have both components. The mural monuments came in a wide range of shapes and designs, and some incorporated the tomb element above which elaborate canopies could be erected (Kemp, 1980). The most common types, however, were set high on the church wall. Again, these varied greatly and whilst the more elaborate could have figurative scenes and be on a grand scale, most were simple panels of text surrounded by architectural features such as pediments and columns, perhaps with a half-figure sculpture of the deceased. More ambitious memorials often had fully carved figures, kneeling or lying at prayer, alone or with their spouses; in many cases children were also depicted at a smaller scale kneeling in prayer (Kemp, 1980; Llewellyn, 1991; 2000; Whinney, 1988). The arrangement of the figures and the architectural settings in which they were placed changed over time and across regional schools, which sprang up as demand grew from the 17th century onwards. By this time monuments were erected not only by major families but by minor gentry and even the most successful of the professional classes (Llewellyn, 2000; Lord, 1986; Sherlock, 1999).

The most common internal monuments were placed on the floor as rectangular ledger slabs, often placed over the actual burial. The great majority of these have been lost usually due to one of three factors. Many were simply worn away by generations of feet, others were replaced by memorials to later burials, and still more in programs of re-flooring the church in the 19th and 20th centuries. Continuing the medieval practice, 16th-century ledgers could have brasses with inscribed images and text in Latin or English (Lewellyn,1991:111–112); surprisingly little research has been undertaken on internal ledger memorials.

Whilst internal memorials may not have been frequent in colonial situations at this period, they would have been known by those emigrating, and were part of the reservoir of funerary forms and decorations which informed choices made by all with a European heritage. Elements could be incorporated or rejected in colonial funerary practice, both reactions being potentially informative of attitudes and aspirations.

5. EXTERNAL MEMORIALS

For all of this early period external memorials were rare, and in most areas only began to appear in any numbers during the 18th century. External memorials in the later Middle Ages were unusual, and few survive *in situ*. The more elaborate were chest tombs (figure 1) but the more common form was the grave slab. These were both forms also found inside churches, and all were to commemorate the elite, though many did not have identifying inscriptions (Burgess, 1963:99–112; Butler, 1964). Many of the medieval forms became popular again in the 19th century as part of the Gothic revival, and as such can be found in many parts of the globe. Late medieval and early historic grave slabs were usually large and

Figure 1. Medieval chest tomb, Fairford, Gloucestershire, England.

rectangular, often with a beveled edge onto which an inscription was placed. They are surprisingly rare, however. Most graves may have been marked previously in the early modern period through piled up earth or stones, by vegetation planting, or simple uninscribed markers, but these would seem to have only lasted over the period of grave visiting, to be swept away within decades. It is therefore during the period up to the 18th century that the concept of more permanent grave marking for the population beyond a small elite was developed.

5.1 Materials

It is widely assumed that many graves were marked with wooden posts or uninscribed rough boulders before the introduction of inscribed memorials, though demonstrating and investigating this proposition has been difficult. Wooden markers are known in England from Essex and Suffolk, but occur most frequently in Sussex (Burgess, 1963:118). The climate is such that wooden examples would rarely have survived outside; indeed the earliest surviving specimens have received considerable conservation and repair. Wooden grave markers are also known in similar forms from North America, such as the grave rails in South Carolina (Crowell and Mackie, 1990:117).

Excavated evidence from Britain to support the postulated use of wooden markers from the 17th century onwards has been hindered by the intensity of overburial in churchyards. However, the presence of small, isolated burial grounds in North America, and the limited extent of overburial, has given the opportunity for more effective examination of the possible use of wooden markers in those contexts. Many farmstead burial grounds in the Chesapeake have evidence of postholes associated with fences, but not with the interments, suggesting that there grave marking was not normally done with wooden posts, though location close to fence lines may have aided grave location by relatives and later grave diggers.

5.2 Forms

5.2.1 External Mural Monuments

In the 17th and early 18th century, monuments may be placed on the exterior walls of churches in Britain, Ireland, and the Continent in the same manner as interior monuments. They are clearly derived from interior memorials, but can be larger and have some characteristics that were later to become features of many headstones. Relatively little research has been undertaken on these monuments, many of which have been lost during later maintenance and rebuilding of the church fabric. The carvers of these memorials often undertook monuments of other kinds, as a study of the Welsh Black Mountain carvers illustrates (Silvester and Pitman, 1998).

In Scotland an important tradition also developed in the 17th century by which substantial monuments were placed against the interior of boundary walls of the graveyard (Willsher, 1985b:22). This was in response to an early and more effective prohibition of internal burial and memorials, and as such frequently led to the creation of external elite monuments with many large architectural features incorporating figural and heraldic sculpture.

5.2.2 Posts, Posts and Rails, and Grave Boards

Many early graves may have just been marked with a timber upright, and an early 18th-century granite monument from Hadley, Massachusetts, is in the shape of a post; it has been suggested that this is a skeuomorph of the wooden memorials in use at that time (Benes, 1975). In other cases it seems that two posts were used, with a rail or board between them, and evidence for their use survives on both sides of the Atlantic. The earliest example noted by Burgess (1963:118) is dated 1658 and is now within the church at Sidlesham, Sussex. This style may have been a regional one in southern England, as it later continued in stone in Sussex alongside other forms of memorial. If the wooden post and rail form had been widespread throughout Britain, one would expect other skeuomorphs to be present, even if the original timber ones had perished, and for such forms to be more widely represented in contemporary churchyard illustrations.

All extant or documented examples of the post and rail seem to have been placed along the length of the grave, with one post at the head and the other at the foot, joined by one or more rails or a board. Text could be painted, inscribed, or carved in false relief, as found on monuments of other materials at this time. The posts could be turned and decorated in the style of contemporary furniture, or with symbols of mortality (Burgess, 1963: 117–118). Three similar memorials survive in South Carolina, and others have been noted in contemporary documents in other parts of the east coast including Georgia, Maryland and New England (Benes, 1975; Crowell and Mackie, 1990; Forman, 1968).

Benes (1975) has suggested that the grave board supported on two posts was transformed into the traditional early headstone and footstone shape. However, post and rail memorials run the length of the grave, so the change in orientation and position of the memorial may not be linked to this form. The popularity of the headstone may have come instead from a combination of improving the unshaped fieldstone placed at the head (and perhaps the foot) of graves to mark them, and the placing in an upright plane the designs and texts from the ledger type memorials. It may be that some graves were surrounded by a fence that could also imitate, in its form, elements of the bed. These designs occur later when the concept of death as sleep was emphasized.

5.2.3 Headstones

The most important external burial monument from the late 17th century onwards was undoubtedly the headstone. Discoid forms (figure 2, left) continue

Figure 2. Left, Discoid headstone, early 18th-century, now lain flat, East Harsley, NorthYorkshire, England. Right, Deeply carved headstone, erected 1696, Stirling, Scotland. Note the range of mortality symbols.

in very small numbers from the Middle Ages (Stocker, 1986) through to the 17th century in England, for example at Lower Swell, Gloucestershire (Burgess 1963: 117), in Derbyshire (Houlbrooke, 1998:363–364) and also in Scotland at Lanark (Willsher, 1985b:19). In Scotland the earliest headstone of a form like later types is dated 1623, and survives at Dunning, Perthshire (Willsher and Hunter, 1978:2); the earliest English examples seem to be from about the same time or slightly later. In New England, the earliest surviving inscribed headstone is from Dorchester, to a Bernard and Joan Capen who respectively died in 1638 and 1653 (Chase and Gabel, 1997:5). It would seem that small numbers of headstones began to be quite widely erected across the English-speaking world by the mid 17th century, with a slow increase thereafter.

During the 17th century, headstones in lowland Scotland were developed with a wide variety of forms, with carving often on both front and back (Wilsher and Hunter, 1978). They continue into the early 18th century and later, with no clear evolutionary development. Often deeply carved, they are usually also very thick (figure 2, right). Similar headstones are also known from County Durham in north-east England, and from north-eastern Ireland (Mytum, 2002d). Many of the stones have moldings on the edges, which are clearly architectural in inspiration, a feature noted in other 17th-century headstones from other parts of England (Burgess, 1963:118). The other source of inspiration was furniture tradition, from which the various 'bedstead' shapes may have derived (Benes, 1975). Many of those involved with carving memorials were also engaged in the building trades as joiners and stone carvers, so these two sources of inspiration are unsurprising.

Most New England stones remained crudely produced till almost the end of the 17th century, and as a result they have attracted relatively little attention from art historians. However, a carver who seems to have been active in the late 17th century in Connecticut has been identified as George Griswold, and he produced some fine headstones and fine lettering, usually with heavy borders and occasionally with decoration such as incised coats of arms (Caulfield, 1991:9–16). The so-called Norwich Ovoid carver has left a small group of late 17th- to early 18th-century simply shaped but undecorated stones in five burial grounds in eastern Connecticut which shows the potential of even less decorative material (Slater, 1996a:5, 46–47). James Stanclift I, who died in 1712, produced well shaped round-topped headstones with capital lettering, and about 60 of his works have been identified in Connecticut and eastern Long Island (Caulfield, 1991:16–23; Levine, 1978; Slater, 1996a:51–54), showing the importance of water transport in the distribution of his products.

The most notable exceptions to the rustic styles of this period are the products of Massachusetts where many fine slate memorials are known from the 17th century. From about 1650 to 1670 stones around Boston had only inscriptions (Tucker, 1993), but thereafter they had decorative features which have attracted study. One of the first of the masons who created these products was the "Charlestown Cutter" of that town in Massachusetts, who worked between 1653 and 1695, and Joseph Lamson who was his apprentice but was operating independently by 1686 (Brown, 1992:169; Tashjian and Tashjian, 1974:63–71; Tucker, 1993).

Virginia stones were largely imported already carved from Britain, because locally available stone was not generally suitable for the purpose, but this ensured that those who did acquire memorials were sufficiently affluent to have ones made of some quality (Crowell and Mackie, 1990:111–113). Markers erected in New York were at first mainly imported from New England, though Welch (1987:3) suggests that one 17th-century headstone carved on both sides was imported from England. Later, from about the 1720s, New Jersey products began to make their appearance, though in New York City they were subsequently challenged by local carvers from 1750, who rapidly became dominant there (Welch, 1987). Established traditions continued west of the Hudson river, however, showing the limited spatial impact of the city products.

The introduction of headstone memorials in any numbers varies greatly from region to region, and only appears at the end of the 18th century in most areas, even those already long settled. The date of inception of the headstone is not as easy to ascertain as might be first imagined, partly because of problems of survival and legibility, and partly by other questions regarding date (see Chapter 9).

5.2.4 Additional Elements

A headstone could be accompanied by additional features, most frequently a footstone that marked the other end of the grave. Flat stones could also be placed over the grave, similar to ledgers (see below), or shaped to reflect either the mound of soil or the shrouded corpse. These elements generally defined and protected the whole of the grave, whilst the headstone carried the identifying text and symbolism. In many burial grounds these additional elements belonging to early grave markers have been cleared or have sunk into the ground, so they are relatively poorly represented. Many footstones have also often been moved.

5.2.5 Ledgers

Stones laid flat on the ground were a popular form of burial monument during this period, and could be used on their own when they would be inscribed, or in combination with headstones when normally left plain and could be shaped into a body or coffin stone. The external ledger had been a popular late medieval burial form for the elite, with finely carved profiles and lettering often around the edge of the stone. Many historic examples do not follow on directly from these, and were often relatively crude. Early Welsh and Scottish ledgers of the late 17th and early 18th century just have simple initials and perhaps a year (Shoesmith, 1980; Willsher, 1985a), though elaborately carved examples with longer texts are known from St. Andrews, Scotland (Fleming, 1931). Gradually more text became added, sometimes to the early stones as they were used by later generations of the family. From the second quarter of the 18th century, many ledger stones had fine carving and lettering.

In New England, ledgers were sometimes called "wolf stones" indicating their role preventing the body being damaged by animals (Chase and Gabel, 1997:5), though many presumed early examples were uninscribed. In most situations it was fear of disturbance by grave digging for later interments that was the greatest impetus, and having a slab which covered the whole of the grave was one of the main purposes of the monument. The ledger stone was the most common surviving memorial form in Tidewater, Virginia (Crowell and Mackie, 1990:131).

Ledger stones occur in a wide range of shapes, though only a few are present in any one region. Rectangular forms are the most common, though others taper towards the foot, or can be of a shouldered, hexagonal shape of a coffin. Most ledgers had a flat upper surface, but some coped examples occur, such as in Scotland (Fleming, 1931; Wilsher, 1985b:19–20). Many ledger stones are plain,

Figure 3. Upper fragment of a ledger stone with deeply carved armorial crest, Killeevan, Co. Monaghan, Ireland.

but there are elaborately carved examples from Scotland (Willsher and Hunter, 1978) and Ireland (McCormick, 1983) where heraldic and mortality symbols are used (figure 3). In most other parts of Europe, any carving was usually mainly heraldic, though mortality symbols were also quite common. Such styles also occur in far-flung trading posts such as those of the Dutch in 17th-century Malacca, Malaysia (Harfield, 1984).

5.2.6 Tombs

Tombs allowed the ledger slab to be raised significantly off the ground, which both prevented it from grassing over and made it more visible in the churchyard. This was achieved either by raising the slab up on pillars, creating table tombs, or on a solid or hollow box which had the advantage of creating a series of vertical faces onto which architectural features, decoration and panels for texts could be placed. This form has been called the chest, box or altar tomb. One region where they were developed was the Cotswolds of southern England (Elliott, 1978; Lees, 1993). Here, narrow and relatively low plain tombs have simply cut inscriptions. These are often on the narrow end panels of the monuments, and occur from the early 17th century. Limited decoration including lozenges, arched panels and fluted pillars, is carved on some monuments. From the 18th century the Cotswold tombs became wider and taller, in proportions that remained the norm throughout the life of this type. Lyre-console ends to monuments occur as a decorative feature that greatly modified their appearance, and was part of a Baroque decorative style. Tombs of the 18th century were plain, but a significant number were highly decorated. Of these the most distinctive were the 'bale' tombs which were taller and narrower monuments with a semi-cylindrical stone set on top (Elliott, 1978:77–80; Mytum, 2000:23). Tombs in North America were rare and relatively plain; the more decorated examples may well have been imported from England.

Early monuments at the Danish enclave of Tranquebar in India had graves, tombs and burial mounds with a Baroque style belonging to the later 17th and early 18th century, though little has been published on these (Nilsson, 1968:136).

5.2.7 Pedestal Monuments

A small number of monuments were constructed to create a vertical impression in the graveyard. In many respects linked to the tombs described above, it was easier to add features such as urns and other sculpture to the top, and around the sides. The shape of early pedestal monuments varied, some being cylindrical, with others based on an oval or square cross section. There were usually several panels for inscriptions around the sides, but rarely on the top. Some of the largest collections of pedestal monuments surviving can be found in the Cotswolds of England (Elliott, 1978), and in the more affluent British settlements of India and south-eastern Asia. They are rare at this period in North America, though become popular later.

5.2.8 Large Monuments

The concept of a large funerary monument was known from the ancient world, and had been occasionally imitated in medieval Christian Europe (Colvin, 1991), but it was actively being used by Moghuls in the Indian subcontinent during the 17th century. The most famous and extravagant example was the Taj Mahal, completed in 1654 (Curl, 1980:138–140), but other less extravagant but still substantial monuments were widespread. The impact of such structures on European traders was considerable, and the most successful could afford emulations of indigenous architectural memorials, with the addition of some Christian and European features (Colvin, 1991). A fine surviving example of such a cemetery is at Surat, used by the East India Company's highest officials at the location of the first British 'factory' in India, where monuments from the mid 17th century survive (Williams, R., 2000).

Many colonial monuments in India were of some size, but were in more European taste. A unique early pyramid style of monument appears in Madras in 1680. The early monuments in Madras have been cleared from around St. Mary's in the Fort, but many of the inscribed slabs have been retained and date from 1652 (Muthiah, 1990:10–12). Many appear from photographs to be from ledgers or chest tombs.

5.3 Decoration and Symbols

Decoration is very varied on memorials of all periods, and this is certainly the case when designs were introduced after the first phase of crudely inscribed markers. There are many regional styles, so even the same motif may be rendered in numerous ways, in part related to the characteristics of the stone and so how it may be carved, but also due to local cultural norms and particular carver's idiosyncrasies. The meanings of symbols (which also vary to a degree between contexts) are discussed in subsequent chapters; here their form and use are briefly considered.

5.3.1 Heraldry

Heraldic designs were common on both internal and higher status external memorials, and on many forms such as ledgers and slabs marking entry to vaults they were extremely prominent. In more complex memorials, several different armorial bearings could be represented, part of a rich genealogy displayed in the monument. On external memorials in Britain, Ireland and in North America, many of the coats of arms were invented, but enforcement over this matter was lax. The College of Arms in London attempted to keep control over the use of arms generally, and had a particular interest in funerals, but there was little that could be done against families that were distant or not concerned with College disapproval (Chase and Gabel, 1997: 505–505; Litten, 1991). Various guilds and trade associations had symbols and regalia, which provided suitable design elements to appear on headstones. These were common in lowland Scotland (Willsher and Hunter, 1978) and northern Ireland (McCormick, 1979), and occur occasionally elsewhere in Britain. Heraldry also occurs on North American monuments all along the eastern states, on ledgers, chest tombs and headstones as well as on the entrances to vaults (Chase and Gabel, 1997:496–604). In Virginia, the use of arms on headstones seems to have peaked in the late 17th century and was rare from the mid 18th century (Kelso, 1992:137–140).

5.3.2 Architectural and Furniture Elements

Many ledgers have architectural designs reminiscent of Jacobean woodwork, such as semi-circular arches supported by columns and capitals. Within this architectural design could be placed the inscribed text, and also symbols. This type of design is found widely on internal ledgers, but also occurs on external memorials, both ledgers and headstones. The hooded moldings on the top of headstones and many other decorative features also are derived from architecture (Burgess, 1963). The relationship between minor funerary art and architecture deserves further attention, particularly within the context of regional patronage and masons' schools.

5.3.3 Folk Art Motifs

Graveyard monuments were carved by those involved in a range of craft activities such as painting, building, or carpentry, and so they were accustomed to using a range of decorative shapes and motifs in other aspects of their work. Such design elements were often incorporated on gravestones, and through study of folk life collections and contemporary illustrations many parallels may be found. These include features from furniture and butter pats (Brears, 1981). In some cases, the motifs have a clear meaning, but many others have no obvious association, though there may have been one for those involved in commissioning and carving the designs (Benes, 1977; Mytum, 2003b). Much folk art on gravestones has few or no parallels in other media surviving today; these numerous commissions for and by the middle orders of society form one of the few sets of specifically commissioned material culture which survives for such groups.

5.3.4 Mortality

Amongst the most notable designs in this period of monumental design were those associated with mortality (see Chapter 5). Popular in all regions and monuments of all types, they varied from full skeletons in anatomical precision to a single schematic skull or long bone, sufficient to signal mortality. The sexton's grave-digging tools and the acroutrements of the funeral such as bell, hearse, shroud or coffin, were all frequently used. Time was another major theme, with hourglasses being particularly popular. In many cases there were combinations of symbols, and these are discussed in more detail in Chapter 8.

6. CONCLUSIONS

During this period most aspects of funerary custom seen into modern times became established, and the combination of coffin burial and memorialization is the norm for the most successful middle classes and above. A pattern of widespread traditions became established across the globe, through the interaction of numerous groups of European extraction. Given the limited social strata involved in memorialization at this phase, only a very small proportion of the European population was represented, and in the colonies indigenous peoples and slaves were hardly reflected at all in this arena of material culture.

Chapter 3

A Maturing Industry

The Mid-18th Century to Early 20th Century

Investment in the funerary process reached its height for the greatest proportion of the population during the later 18th and 19th centuries. As the expanding middle classes and, increasingly, the working classes, participated actively in funerary and commemorative display, the industry reached its maximum of both quantity and complexity (Curl, 1972; Litten, 1991; Morley, 1971; Sloane, 1991).

The material trappings of the funeral had all been present earlier: the invitations, mourning clothes, the coffin and the funeral procession, but these were much developed. Within an expanding capitalist market, mass produced goods allowed many more families to participate in these rituals. Undoubtedly much of the expenditure was sometimes regretted by those paying for it, and this is reflected in the cartoons and some literary satire (Jalland, 1999:243–244; Sloane, 1991:173) but social conventions and limited choices constrained peoples' options at a time when grief and uncertainty reduced individuals' power and resolve. However, many no doubt gained great solace from the socially respectable funeral of a deceased relative, something noted in much contemporary literature (Douglas, 1975; 1977:200–226; Jalland, 1996).

The items of material culture necessary to support the mortuary rituals of the Victorian era were many and varied (Morley, 1971). These all provide information in their own right but together show the rich pattern of mourning and commemoration which provided an almost constant thread through the lives of all at that time. The rituals were particularly important to the middle and upper classes, and were displayed intensely in the domestic arena as well as at the burial site. As a consequence, the mourning rituals affected women to a greater degree than men. The scale of the funeral and subsequent mourning rituals varied according to wealth and prestige, and numerous catalogs and guides to etiquette could aid the client. The increasingly professional undertaking industry grew in a remarkably similar manner in Britain and Europe and the more densely colonized possessions, and in North America (Habenstein and Lamars, 1981; Kselman, 1993; Litten, 1991; 1997; 1998).

The materiality of the mortuary and commemorative process during this period can be broken down into three elements. First there were the events leading up to and around the funeral; then there were the phases of mourning; and finally there were long-term commemorative practices. Most archaeological research has concentrated on elements of the funeral through excavation of burials, and commemoration through memorials. However, it is worth setting these in a wider material context.

1. THE FUNERAL

Following death, most individuals were laid out in the family home, and often dressed in grave clothes; this had been the case in the earlier period, but only documentary evidence is available in any quantity (Gittings, 1984). From the 18th century onwards, examples of clothing have survived not only in museum collections but also recovered from excavation. Significant numbers of clothing items were recorded at Spitalfields, London (Janaway, 1993; 1998), but they survive elsewhere in unpredictable ways in coffined burials within churches or other dry conditions. A wide range of materials were used for the clothes (Litten, 1991:81), which did not have to be well finished as they would suffer no wear.

1.1 The Coffin and Preparation of the Body

The body was laid out in the coffin, usually lined with fabric, and was available for the bereaved to view. The social importance of this phase of mourning and loss was most important, and the corpse had to appear at rest. This involved manipulating the body, tying it if required, and using mattresses, pillows and shrouds as necessary. Sawdust was often used in the base of the coffin to absorb any liquid which might emanate from the corpse, and to help keep the body still during its movement to church and then grave. This, and fragments of vegetation to hide any odors of decomposition, may be found in excavation and were noted at Spitalfields, London (Reeve and Adams, 1993) and North Dalton, Yorkshire (Mytum, 1988). The evidence from Spitalfields has been reviewed to consider the many ritual facets of the funeral process that are visible archaeologically (Cox, 1998a).

Slave funerals in the Caribbean involved the preparation of the body in clothing. This was followed by a procession with, by this period, the deceased carried or on a hearse, followed by mourners in pairs. Some slaves were buried in coffins, but many were not.

Preparation of the body for burial during the 19th century changed most dramatically in North America where the trend towards embalming was at its strongest. Begun in numbers during the Civil War, this developed as part of the

Figure 4. Fragment of coffin with fabric held in position by double lines of upholstery pins. Note the coffin grip (handle) and grip plate. Kellington, North Yorkshire, England.

funeral director's art (Sloane, 1991:174), and led to even more manipulation of the appearance of the deceased prior to burial. The use of glass viewing windows in North American coffins from the middle of the 19th century (Bell, 1990:58) reflects this concern and also the ability to provide a body appearing at sleep. This became a common feature on North American coffins, but it was not used in Britain.

Coffins continued to be produced mainly in the flat-lidded six-sided single-break form (Litten, 1991). Various local woods were used for construction, using panel pins or nails. In more affluent areas the coffins were then covered with fabric held in place with upholstery pins; black was the most common, but dark red, green, blue and purple were available, and white was used for children. The coffin was then furnished with handles (grips) and lid motifs, usually including a breastplate on which an inscribed identifying text (figures 4, 5). The pattern of multiple shells available in the earlier period continued for the wealthier clientele. Iron coffins, including the Fisk metallic coffin, were popular in the middle to later 19th century in North America, and have been recovered archaeologically (Habenstein and Lamars, 1981; Owsley, 1997:7–8). There are also known from documentary sources in Britain and Ireland (O'Neill, 1965). Wooden caskets, rectangular in shape and more substantial than coffins, became popular in North America but less so elsewhere, though occasional rectangular coffins are found archaeologically in many regions (figure 5, right). Some slaves were given coffins by their owners (Handler and Lange, 1978:191–192) and the use of coffins by

Figure 5. Left, Flat-lidded six-sided coffin in a brick-lined grave. Right, Roughly rectangular coffin with strips of lace (decorative metal edging) and shield-shaped breastplate in a brick grave shaft. Both from Kellington, North Yorkshire, England.

African Americans increased considerably during the 19th century (Courtaud et al., 1999; Jamieson 1995:53; Watters 1994).

Litten (1985) has provided a chronology of the coffin furniture for England. Several materials were used on the 18th century fittings, notably lead, copper (sometimes gilt), brass or tin-dipped iron (called silver by the trade, and tinplate today). The last could be painted over in black or cream. From 1769, a stamping method was invented which allowed more variety at little cost using tinplate (Church and Smith, 1966:621). During the latter part of the 18th century, large numbers of metal elements could be applied to the coffin grips: (handles) and grip-plates (figure 4), pins (upholstery pins to hold in place the fabric covering the coffin), lid motifs, escutcheons and breast plates on which any commemorative details would be incised (Litten, 1985). The coffin could be further decorated by visible hinges, and stamped metal termed lace could be used to edge the coffin instead of pins (figure 5, right). During the early 19th century fabric was less often used to cover coffins which were now usually french polished and had fewer fittings. By the end of the 19th century, often only the grips and breast plate was chosen. This same pattern of increasing elaboration followed by a move to greater simplicity is noted in North America, where many individual sites have been excavated with such fittings (see Bell, 1990; 1994 for references). Coffin fittings for childrens' graves were similar in style to those of adults, but fewer in number and sometimes size because the coffins were smaller (McKillop, 1995).

1.2 Increasing Commercialization

Some aspects of earlier funerals continued, but the popularity of night funerals waned and they were rare after the middle of the 18th century. The funeral procession, with its paraphernalia, developed for a widening clientele. The funeral director became the purveyor of a wide range of facilities, many provided directly and some subcontracted out to others. These services included the provision of the coffin and its fitting, the rent of the hearse and the hire of paid mourners, the arrangement of printed invitation cards, gifts to be distributed, mourning clothing, and funeral drapes for the home (Habenstein and Lamars, 1981; Jumonville, 1993; Litten, 1991). From the highest in the land to poor laborers, a funeral package could be identified which matched pocket and social status. Only the paupers did not get such a service, and as far as possible working class people tried to avoid having to suffer the indignity of a pauper burial. The poor were encouraged from the later 18th century to contribute to burial clubs, thus ensuring an adequate and respectable funeral; these became particularly important in the 19th century (Litten, 1991:164–65). Many North American fraternities, associations and unions had funeral support, and indeed some had their own burial plots in cemeteries (Florence, 1997).

At slave burial grounds various graveside rituals were undertaken; little detail is known of these, though some led to the provision of grave goods (Handler and Lange, 1988:184–203). Some special burials have been noted that may relate to those with ritual knowledge (Handler, 1997). Orientation of slave graves became standardized, despite varying traditions in Africa, to east-west; although the head was to the west, a significant minority had head to the east (Jamieson 1995:52).

In America by the 1880s, more people were dying in hospitals than at home, though it was still the tradition to transfer of the body back to the family residence and have it laid out there. The role of the funeral director was growing, in arranging this and much of the subsequent ritual; it was well into the 20th century, however, before this transition became complete and widespread (Habenstein and Lamars, 1981:257–285; Sloane, 1991:119–120). In all parts of the industrialized world, from France (Kselman, 1993) to America (Habenstein and Lamars, 1981), from Britain (Howarth, 1997; Litten, 1991) to Australia (Zelinka, 1991:68–71), similar patterns of commercialization can be discerned, with remarkably common features. This is not to deny national, regional and local variations, often associated with ethnic traditions and religious emphases, but the similarities are worthy of note.

1.3 Popular Fears Regarding Burial

Two fears developed during the later 18th and 19th century which had an effect on the material culture associated with the funeral. The first of these was of being buried alive, and this was assuaged by careful inspection and attendance upon the body as it rested in the home, sometimes for many days. In parts of Europe the bodies were left in public mortuaries with attendants who could be summoned by those not actually deceased (Litten, 1991:166). The other fear in Britain, at least, was that the 'resurrection men' would come and exhume the body for medical dissection (Richardson, 1988). To prevent this, various tactics were employed to protect the body, or render it useless for study, and the material manifestations of attitudes to the disturbance of the dead are discussed further in Chapter 8.

2. MOURNING

Mourning was demonstrated most effectively through dress, but behavior was also curtailed. On both these counts, only those with the resources and the facilities to adjust activities could take part in most of these rituals. Many people must have merely worn some element of black, and the poorest could have continued to manage as normal. For the middle and upper classes, however, a scale of mourning clothing was set out, though this changed over time and in social context (Cunnington and Lucas, 1972; Taylor, 1983). Crape became the standard material associated with mourning during the 19th century, and was used extensively for drapes and for clothing. The type of clothes worn, and the length of time, depended on the relationship with the deceased, with widows suffering the longest period in mourning. Advice in 1880 was that it was two years before half mourning—which allowed other colors such as gray and violet—could be undertaken. This lasted for a further six months before normal clothing could be worn. For other relations a year or six months could be sufficient, and gradually rules were relaxed regarding the period in mourning, though there were many vested interests in the textile, fashion and funerary industries who wished to maintain such an exacting and expensive regime (Morley, 1971:68–79). Clearly, settlement sites should on excavation produce numbers of items such as buttons and jewelry that reflect the mourning costumes.

2.1 Mourning Jewelry

Mourning jewelry was also produced to augment the costume, and could be in a range of styles (Bury, 1985). Classical revival motifs of mourning women, tombs and urns were popular, as were flowers, silhouettes, and by the later 19th century, Gothic revival designs. Locks of hair could be incorporated into jewelry, or hair could be used on its own to create items such as pendants and earrings (Morley 1971:66, pl.91; Pike, 1980). Jet, a hard form of coal or similar material, was very popular and could be carved into elaborate designs and with a dull or shiny finish (Morley, 1971:66). Black glass, cut to create an effect more like precious stone, was termed French jet, and was also widely worn. These items may be found on settlement excavations and in refuse dumps.

2.2 Other Mourning Material Culture

Other items of material culture that related to mourning included memorial cards, similar to those produced as invitations to funerals, which could be kept on the mantelpiece. Photographs of the deceased, particularly infants, were specially taken by professional photographers to act as *momento mori* (Ruby, 1984). Children and women could produce mourning samplers (Clabburn, 1998; Schorsch, 1979), and other items of needlework including extremely competent silk embroidery, concentrated on memorial themes. The shared use of motifs, sentiments and texts on gravestones and needlework has been highlighted by Gabel (2002). Other forms of commemoration should also be considered, from the lists of deceased members written in the family Bible through to specially purchased sets of mourning ceramics (Morley, 1971). Many memorial items within the home could be used by Catholics as a focus for prayer, to lessen the time spent by the soul of the departed relative in Purgatory. The location of memorabilia within the home would relate to the type of remembrance required, whether in quiet personal contemplation in the bedroom or more public recognition in the parlor.

3. COMMEMORATION

Photographs became increasingly significant in relation to long-term commemoration, though they could also be used prominently in mourning (Ruby, 1984; 1995). Photographs could be taken of the deceased, but for most adults pictures of the individual in life were preferred. Memorial or funeral cards could be framed and mounted on the wall as a permanent remembrance. Many national figures were commemorated in life, but equally so in death, with photographs and inexpensive printed pictures. All types of leader could be so represented, from great preachers and missionaries to politicians, royalty, explorers or the military. An American example would be the material produced on George Washington's death (Gabel, 2002; Schorsch, 1979), or in Britain that linked to the death and funeral of the Duke of Wellington or Lord Nelson (Litten, 1991). Such individuals were also commemorated in many ceramic souvenirs (Morley, 1971). Tragic events such as colliery accidents were also remembered through such products, which could also be sold to raise money for those affected by such disasters. As research is carried out in the accessible and surviving arenas of burial grounds, the range of alternative or additional commemorative strategies therefore needs to be remembered.

Acts of commemoration within burial grounds took place on the anniversary of deaths, and in many regions on All Saints' Day. Graves would be tidied, and flowers put in place. This is considered further in Chapter 8. Other activities could take place elsewhere on anniversaries, though these were of a personal nature and may be best studied through diaries and letters as they usually left no physical trace.

4. BURIAL GROUNDS AND CEMETERIES

The traditional location for archaeological evidence for mortuary customs are burial grounds, and these do indeed provide enormous potential for this period. The evidence can and does relate to many wider social and cultural issues beyond that of deathways, so the aspects of use of space and material culture should where possible be integrated with wider trends within a particular community, cultural group or period of time.

4.1 Location

The choice in burial location widened during this period from the traditional ones, largely controlled by the church, to a range of settings, often under secular control. Greater choice in location led to the development of landscapes and memorials which can be linked to shifts in architectural and landscape fashions, though changing attitudes to the body and remembrance were also important.

4.1.1 Rural Locations in Britain and Europe

Most existing traditional rural burial locations in Britain and mainland Europe continued to be used throughout this period, though with the expansion of nonconformist sects and greater tolerance of their separate funerary arrangements, the numbers of burial grounds increased considerably. The same pattern of continuity combined with denominational diversification can be seen in the colonies, with the establishment of family plots on farmsteads, separate slave burial grounds on plantations, churchyard and chapel burial grounds. Some of the new burial areas were adjacent to places of worship, but others were on separate plots because of availability of land. This dislocation of burial and commemoration from the church and its regular Sunday worship was a precursor to a pattern created by separate civil cemeteries, first only established in urban contexts but later communal rural ones were also founded.

Ireland had a largely Catholic population, but Anglicanism was the official religion. Many new Anglican churches were constructed in the 18th century, with associated graveyards, but much rural Catholic burial continued to take place inside and around the ruined medieval churches. The Anglican churchyards were also dominated by Catholic burials and, as a Catholic middle class emerged, their memorials became numerous (Mytum and Evans, 2002). Internal monuments, however, largely remained the preserve of Anglicans. Unbaptized infants were interred in separate burial grounds, *killeen*, some of which were also on earlier religious sites.

In northern and western Europe, many churches have associated churchyards for burial not dissimilar to those in Britain, though many have few older stones because of subsequent reuse of the plots. In the Mediterranean, burial was largely in communal village cemeteries, close to the settlement. However, some wealthy families in Greece had burial chapels on their estates (Bennett, 1994:123), a pattern noted anecdotally elsewhere in the countries around the northern shores of the Mediterranean.

4.1.2 Rural Locations in Colonial Contexts

As colonial exploration spread across the globe, an ever-expanding frontier meant that some of the population were living and dying in regions without the infrastructure found in the homeland, nor with a great density of European population. In such situations, isolated *ad hoc* burial was frequent, usually with little marking the location of graves (Mytum, 2003a; Sloane, 1991:14). The fear of an isolated death meant that such sacrifices were often imbued with particular power in paintings and literature, a trend found in Britain, North America and Australia (Fitzpatrick, 1997; Llewellyn, 1991:34–36).

In maturing colonial situations such as New England, some different trends can be noted. A shift of burial areas from separate and relatively isolated plots to locations adjacent to the places of worship was particularly strong in the mid 18th century (Brooke, 1988:465). However, the strong presence of a variety of sects in a parish, such as Baptists and Quakers as well as the orthodox Congregationalists in parts of Worcester County, produced a fragmented, scattered pattern of interment in sectarian burial grounds, similar to that seen in Britain.

In America, rural farmstead burial continued in those long-settled regions, and expanded to the north, across the mid west, and to the west coast. It also developed on plantations in the south (Stokes, 1991:178) and the Caribbean (Handler and Lange, 1978; Jamieson 1995). Small rural cemeteries are frequent, some associated with particular churches, others for whole communities. These small cemeteries were founded in increasing numbers after the Civil War. Most reports on excavations at these burial grounds are in the regional 'gray' literature (see Bell, 1994 for many examples), but the potential can be seen in some that have reached a wider audience (Burnston, 1997; Maish et al., 1997). Small areas of land were marked out often in an impermanent manner and so have been lost, but are frequently encountered in development schemes, as was the case with the Walton family cemetery at Griswold, Connecticut (Bellantoni et al., 1997).

Long-lasting communal burial areas under some form of secular control often became established as population densities rose (Sloane, 1991:14–15), and distinct regional styles developed. Some of the most notable of these include the Upland South cemeteries which have attracted some sustained research (Jeanne, 1969; 1978; 1989). They tend to be placed on hills, away from places of worship which were often only built at a later stage of settlement after the burial ground had already been established and used. The pattern of evolution in North Carolina has also been documented (Little, 1998). Survey in part of Illinois has identified small cemeteries from the first half of the 19th century with under 10 graves, few with markers and now abandoned, and small family plots of up to 20 graves, founded after about 1860, but now also disused. These were usually on raised round and marked by cedar or black locust trees. Cemeteries founded in the late 19th century were usually at a focus of rural activity such as a church, road junction or other feature, and continue in use to the present (Price, 1966). This pattern is probably applicable in general terms to many areas of North America. The *camposantos* of the south-west have also been examined (Warren, 1987), though most remains come from late in this period, and so a more detailed chronology has yet to be developed. The various forms of burial ground in Texas have been well described by Jordan (1982).

4.1.3 War Cemeteries

War cemeteries were necessary in many parts of the world, but have only been studied in any detail in America. Most research has concentrated on the battle sites in order to understand the battle itself, and the burials afterwards have not been the main center of attention. In the famous case of the Little Big Horn, much archaeological investigation has indeed focused on the events, but the burials have also been explicitly considered (Fox, 1993:73–77; Scott et al., 1989). Napoleonic cemeteries are now being studied as they are revealed by development in eastern Europe, though most emphasis has been on the pathology.

Surface survey of war cemeteries has been widespread, though the literature has tended to concentrate on the main architectural features. American Civil War cemeteries were the first to be constructed which were regularly laid out, and were termed National Cemeteries. They could include memorials for individual soldiers; at first these were of wood, but from the 1870s small granite and marble monuments were erected. Such sites had carefully designed layouts and created appropriate, democratic settings for commemoration. National Cemeteries have continued in use both for war dead and veterans who are entitled to be buried there if they wish (MacCloskey, 1968; Sloane, 1991:113–115).

4.1.4 Urban Locations

From the second quarter of the 18th century, parish crypts began to be constructed beneath urban churches to satisfy middle class demand as churchyards became increasingly crowded. Examples include that at Spitalfields, London which has been subject to archaeological investigation (Adams and Reeves, 1987; Reeve and Adams, 1993), and those of churches in New York such as the African Zion Methodist, the Episcopal, and Trinity churches known from contemporary accounts (Sloane, 1991:37–39). These rapidly became filled, and their management became problematic. The internal family vaults in parish churches continued to be used, and some private vaults were constructed off parish vaults.

African-American churches in America suffered the same problems of overcrowding experienced elsewhere, exacerbated by limited resources. The limited financial security of such congregations often led to movement of places of worship and abandonment of burial grounds, which were then redeveloped, and several in Philadelphia have been examined (Crist et al., 1997; Parrington, 1987). It is noteworthy that a substantial number of churches of all denominations, and their associated burial grounds, have also been demolished and built over in Britain and Europe. Recently examined examples include the Anglican Cross Bones burial Ground, Southwark, London (Brickley et. al., 1999), and the Quaker burial ground at Kingston-upon-Thames, Greater London (Bashford and Pollard, 1998).

During the 19th century there was a great deal of contemporary discussion about the noxious nature of graveyards, and also the unsavory conditions within churches with so many burials taking place there (Curl, 1972). The closure of city churchyards became necessary as urban populations rose and overwhelmed the limited space in churches and churchyards. Alternative arrangements were thus required for the large and increasing numbers of burials required each year in towns and cities across the globe. The solution favoured was that of large, secular cemeteries, though the ownership, management, style and other function varied greatly (Table 2). The development of cemeteries was inhibited by vested interests, usually of churches but sometimes others, but they were in the end the widely adopted solution across the globe. In Britain, joint stock companies, local authorities and even some religious institutions developed sites across the country (Brooks, 1989; Curl, 1972; 2001; Mytum, 1989; Rugg, 1998). Most European countries favored local government control (Ariès, 1981; Brown, 1973; Etlin, 1984; Goody and Poppi, 1994; Kselman, 1993; Quer, 1989; Ragon, 1983).

In America local government and private companies set up most cemeteries, though some secular societies and religious institutions established them also (Florence, 1997; Masson, 1993; Sledge, 1993; 2002; Sloane, 1991). In colonial cities it was government that took the lead, with administrators at times overcoming or ignoring denominational and ethnic disagreements to produce arrangements which suited the government view (Christopher, 1995; Mytum, 1993; Weston,

Table 2. Types of early coffin (based on Litten, 1985, but see also Butler and Morris, 1994; Litten, 1991; 1999 for examples and illustrations)

Form	Dating	Description
Anthropomorphic	to c. 1670	Lead, shaped to fit corpse. Lid with anatomical Shell features, lead inscription shield normally above the abdomen. Could be placed in a rectangular chest with iron bands or a single-break flat-lidded shell.
Rectangular flat-inscription lidded lead shell	to c. 1650	Has inner elm coffin. Small rectangular lead plate above position of the deceased's chin.
Tapered flat-lidded inscription lead shell	to c. 1650	Has inner elm coffin. Small rectangular lead plate above position of the deceased's chin.
Gable-lidded single-break shell	to c. 1650	An attempted revival in the 19th c was unsuccessful. Can be in lead-lapped wood or elm.
Fittings for coffins to c. 1650 were wrought iron grips, lead breastplate and pins in a panelling pattern which held the covering of black velvet in place.		
Single-shell	c. 1650 –	Single-break, flat-lidded.
Single-shell double lid	c. 1650 –	Single-break, flat-lidded.
Double wooden shell	c. 1650 –	Single-break, flat-lidded.
Triple shell	c. 1650 –	Single-break, flat-lidded. Lead-lapped inner wooden shell with outer wooden case.
Fittings for coffins from c. 1650 were stamped iron breast plates and grip plates, with larger numbers of pins used to hold the covering of velvet or baize. The normal fabric colour was black till the early 18th c when scarlet became popular. By the late 17th century, initials, age and year of death could be marked out using upholstery pins on the coffin lid.		

1989; Yeoh and Hui, 1995; Zelinka, 1991). Some cemeteries could be sited some distance from the city, and the Brookwood cemetery, Woking, England (Nash 2000) and Rookwood, Sydney, Australia (Singleton and Kay 1989) were both reached by railway, with special funeral trains laid on as required.

The development of new settlements and the transformation of others with the industrial revolution led to the creation of many new urban burial areas associated with industries such as textiles which required water power, or those which extracted coal, iron or other natural resources. At planned settlements the same degree of control could be exercised by the developer over places of worship as over factories and housing, a feature found on both sides of the Atlantic. This sometimes also included burial provision (McGuire, 1988; 1991).

In colonial contexts, segregation was often employed, with distinct newly designed cemeteries for different racial groups, as well as different denominations. For example, the Anglican churchyard around St. John's church Wamers, South Africa, was only for Europeans, and a separate burial ground was laid out for native Africans (Christopher, 1995:41). As racial segregation intensified in the late 19th and early 20th century, separate cemeteries were established associated with the ethnically defined suburbs that were being constructed, a process found over the whole of South Africa (Christopher, 1995). Slaves, presumably baptized, were buried in Barbadian churchyards though this was rare until the 19th century, when there was a greater interest in converting slaves to Christianity and numbers of churchyard interments rose (Handler and Lange, 1978:174–177).

The situation regarding racial separation varied across America. Cost was the only factor in most of the northeast, but there were planned segregated sections in the mid west, and completely separate cemeteries in much of the south (Jordan, 1982; Little, 1998; Sloane, 1991:83; Stokes, 1991). A detailed study of Louisiana cemeteries has revealed clear patterning of distinct burial places based on denomination and race (Nakagawa, 1990; 1994). Here, European American Catholics, European American Protestants, and African American Protestants could all be distinguished as having sufficiently distinctive cultural preferences; these tend to have distinct regional distributions because of the composition of the population within different parts of the state.

4.2 Intra-site Spatial Arrangement

The study of intra-site patterning can be carried out rarely with a study of both surface commemorative markers and the interments within the burial ground. Usually, excavations are on sites with few markers, or these have already been cleared. Studies of surface memorials are the easiest to conduct, but it should be remembered that many interments may not be marked, or that commemoration may take place on a monument which was not at the location of the burial. Nevertheless, intra-site spatial analysis offers many opportunities for research.

4.2.1 British and European Rural Graveyards

European rural graveyards are not generally amenable to intra-site studies for this period as most memorials of any antiquity have been removed in the process of grave plot reuse, though one Danish study has been published (Nielsen, 1997). In Britain and Ireland evidence is more plentiful, but even here there has been limited research. The normal pattern would seem to be one of continued use of the area south of the church, but a wider area becoming used for burial to the south, east and west. As the land began to fill up with memorials, the northern part and peripheral areas were used, and plots not marked with memorials were reused and infilled with later memorials (Mytum, 1996; Shoesmith, 1980).

Rural graveyards often became perceived as full, and an additional piece of land would be taken into use for burial; some graveyards have had several additions. Extensions to churchyards, and newly established rural burial grounds of other denominations, were laid out in rows with efficient use of the space. With no anticipated reuse of burial plots, the ground was to be made to last as long as possible, to prevent or defer the need for a further extension. The planning of these later burial grounds or extensions was often therefore very different to the earlier churchyards, a situation further emphasized by the much higher proportion of plots marked with memorials (Mytum, 1996, 2003d).

British rural churchyards were planted with specimen trees and bushes to give an appropriate effect, and similar attempts were made in colonial settings. The selection of vegetation was in part related to the funerary and memorial context, but was also influenced by the wider fashions in garden design and planting (Gough, 1996). In the 18th century, yew was popular, and was often clipped to shape. Where this has continued some feel for the setting can still be felt, as at Painswick, Gloucestershire (figure 56), but in other places the yews have been neglected and have grown into trees. Other types of conifer and evergreens became popular during the 19th century, though native deciduous species were still often found. Where initial designs for colonial cemeteries and churchyards may have initially favored British or European species, those that did not thrive were replaced by indigenous species, creating distinctive funerary flora in various parts of the world (Burke and Betteridge, 1989).

4.2.2 North American Rural Burial Grounds

North American rural family burial grounds tend to have regular rows of burials, set within a plot defined by fences, walls or railings. Markers were often simple, and did not always survive for long, but the limited evidence of intercutting on excavated sites suggests that the location of burials was either known to family members or could be deduced on surface evidence by new owners. Racial segregation occurred within family burial grounds, and African Americans rarely had grave markers when buried in the graveyards of their owners or employers. Many rural family graveyards have a large memorial for a dynastic founder or a particularly successful ancestor. These monuments may have been erected soon after death or may be a much later commemorative expression. The Thomas Jefferson monument at Monticello, Virginia, has been replaced with a larger monument, but it always was a tall obelisk, with other graves marked with much smaller stones (McDowell and Meyer, 1994:137–140). This pattern was repeated widely, and also became the form on family plots in cemeteries (e.g. McDowell and Meyer, 1994:141). The social implications of this phenomenon are discussed in Chapter 6.

Rural communal burial grounds were often not large, but could be extended as they became full. Most published research has focused on rural graveyards in the south. Stokes (1991) notes that burial locations tend to be randomly chosen within the burial ground in Florida, partly because of the amount of vegetation and presumably no effective central control over such sites, though east-west orientation is always maintained. Nevertheless, some deliberate planting of cemetery trees can be noted, including cedars and other evergreens at both European and African American burial grounds (Stokes, 1991:189). Some Florida sites show the scraped earth maintenance regime of the upland south tradition investigated by Jeanne (1989). He suggests three phases of development of such burial grounds, two of which relate to this period. Burial starts with a pioneer phase with the whole area scraped of vegetation, and earth-mounded graves. Pine and oak may be planted, but the most common species placed in such locations is even termed the 'cemetery tree', the eastern red cedar (Jeanne, 1989:114). Wooden grave shelters occur only on European cemeteries and although this feature has been linked to native American funerary traditions, Jeanne considers that it is more likely derived from structures such as the British lychgate found at the entrance to many churchyards. In the later transitional phase of Upland South cemetery development, Jeanne proposes that newly enclosed burial areas were not scraped, though the burial plots were. At this phase grave markers became more frequent, and the edges of the plots may also be demarcated, but the use of the grave shelter waned. Many different species of shrubs, including flowering ones, were introduced (Jeanne, 1989:116).

Hispanic cemeteries in New Mexico were inspired by the traditions of traders who arrived from the east in the 1820s, and were on a rough grid arrangement, with scattered burials. The rural *camposanto* graveyard began to acquire not only a few stone carved headstones but larger numbers of wooden crosses, and from the 1880s a range of shapes and designs, often painted, emerged. Many plots at this time were marked by wooden picket fences, though they were rarely surviving by the 1920s (Warren, 1987).

4.2.3 Urban Churchyards and Cemeteries

Urban churchyards in many countries became fossilized once burial was suspended, and were often reduced in size as demand for land increased and streets were widened. As memory of the deceased faded, and grave visiting stopped, some urban churchyards were partially or even totally cleared, a process which has continued into the late 20th century (Mytum et al., 1994). In those parts of the churchyards that survived many memorials remained in place, however, and family owned vaults and shafts continued to be used for burial in some cases. In America, many of the older burial grounds in small towns such as those of New England were left undisturbed once full, and new cemeteries established on the peripheries of the settlements. It is in the older burial grounds that so many of the famous 17th- and 18th-century mortuary monuments survive. In some cases the African American population, whether slave or free, had a small area on the edge of the burial ground set aside for their use. At Newport, Rhode Island, this has survived and continued to have burials until the middle of the 20th century, and has attracted some academic attention (Garman, 1994; Tashjian and Tashjian, 1989:163). Very many burial grounds were built over, especially in rapidly expanding cities such as New York and Philadelphia, and may become available for excavation when further redevelopment takes place (Crist et al., 1997).

In Britain, Europe and North America some early cemeteries were organized on a regular grid layout (Curl, 1994), reflecting the practical need and philosophical inclination that many of the cemetery founders had in replacing the often chaotic and overcrowded churchyards. This regular plan was also followed in newly founded cities in the frontier, as with Oregon cemeteries which only became softened with some circular elements and fountains in the late 19th century (Francaviglia, 1971:506–507).

Many cemeteries in Europe and the developed regions of the colonized world were given a more sophisticated structure than the simple grid form. This appealed to the sensitivities of their owners and their financially if not numerically most important middle class clients. Cemetery research has especially concentrated on the development of the Garden or Rural Cemetery concept, both in Britain (Curl, 1984; 1994; Penny, 1974) and America (Bender, 1974; Linden-Ward, 1989a; Reps, 1965; Rotundo, 1974; 1984). This was a style of cemetery design which favored a picturesque and romantic environment, landscaped and planted appropriately (figure 6). Within this idyllic setting a business-like management structure provided plots to individuals, families and associations in which to bury and on which monuments could be placed. It was found throughout the colonial world and has

Figure 6. Monuments set within a carefully designed and planted landscape, Mount Auburn, Boston, Massachusetts. Note the family grouping of memorials, some within kerbed plots.

been studied in Australia (Burke and Betteridge, 1989). Much attention has been paid by researchers to the design and aesthetics of cemetery planning, but much less on the way in which the cemeteries became filled. This would reveal much about changing attitudes to burial and memorialization, and the resultant changes in physical appearance of the cemetery that these processes caused. This again is difficult in much of Europe because of survival, but could be achieved in Britain and its colonies, Ireland, and America. The current literature reflects the large influence of art, garden and landscape historians on this research area, rather than archaeologists.

The filling of rural style cemeteries and the vast array of monuments within them led to a reaction in North America which led to the development of the landscape-lawn cemetery (Sloane, 1991). This change involved a greater management control over family plots and their appearance. This reduced grounds maintenance costs and created a more uniform aesthetic experience. Starting at Spring Grove, Cincinnati, which had been founded as a rural cemetery (Linden-Ward and Sloane, 1985; Linden-Ward and Ward, 1985), this style became the dominant choice for new cemeteries in the second half of the 19th century. Changes in existing cemeteries, as at Spring Grove, were made when objections of plot owners could be overcome. Sloane (1991:110) notes the impact of the invention of the lawn mower in England in 1830 had profound implications for management of cemeteries. Now, vast swards of short grass could be created and maintained throughout the year. The softer rolling contours of the lawn cemetery could now be achieved.

Many cemeteries around the world became more like the North American lawn cemetery, though often without the same control over plot monuments. The more exotic and romantic features of the rural cemetery were abandoned, and a more functional park-like landscape created, often further emphasized by the use of seasonal bedding plants. Where cemeteries grew in size, the original area may be of the rural style, with later additions of the regional variety of the lawn type.

European cemeteries have received some architectural research, mainly concerning the original planning of the sites, and the major monuments subsequently erected within them (Colvin, 1991; Malachovicz, 1996; Quer, 1989). Only France, however, has received continued research, and this has included studies which include smaller towns as well as Paris (Etlin 1984; Kselman, 1993; McManners, 1981). There has also been some significant anthropological study of Mediterranean deathways, but with the inevitable concentration on 20th century attitudes, and less on the early development of burial space and monuments. Nevertheless, Bennett's study in Lehonia (1994:122) noted that most burials were placed in the traditional village cemetery at this period without any spatial differentiation according to status, a situation not replicated in the modern cemetery.

With the establishment of cemeteries came complex sets of charges for external graves of particular types (Barnard, 1990:200–203), and zoning by ability to

Figure 7. Substantial family mausolea in a desirable location within the Père Lachaise cemetery, Paris, France. Note the densely packed appearance.

pay was commonplace, continuing traditions already long established with regard to church and churchyard burial in Britain and the Continent. This began with Père Lachaise in Paris (figure 7), where plots were graded in price, following on the tradition of differential pricing based on location and duration of interment in the church-controlled burial locations (Kselman, 1993). In some cemeteries, certain prominent plots were only available provided a suitably ostentatious monument were to be erected upon it, thus ensuring the aesthetic and social standing of the cemetery as a whole.

The arrangement of zones for different Christian denominations and other religions was also deemed necessary. The amount of space and its location within the cemetery reflected the numbers within the community but also their overall social standing. These categories and apportionments were adjusted over time, with cemetery extensions designed to create more space for those most rapidly filling their allocation. In some American cities, cemeteries tended to be created on a religious basis, often augmenting a multi-denominational cemetery. For example in Mobile, Alabama, the main cemetery that replaced the Church Street graveyard was the Magnolia cemetery, owned by the city (Sledge, 2002). This, like its predecessor had sections for different religions and denominations, but was subsequently augmented by the Old Catholic cemetery (Sledge, 1993) and two Jewish cemeteries. Non-denominational privately run cemeteries were established during the 20th century as Magnolia cemetery became almost full (Sledge, 2002). In southern Europe, a few Protestant cemeteries were established for burial of ex-patriots and travelers as well as any local religious minorities. Those at Rome and Florence maintained clear distinctions from the local norms in their monument choice, layout, and management practice (Freeman, 1993; Nylander et al., 1989; Rahtz, 1987).

The perceived need for racial segregation within the burial site was a significant element in the planning and design of cemeteries in some American and British colonial contexts. For example, New End Cemetery in Port Elizabeth, South Africa, was established in 1863. Here, Wesleyans and Wesleyan natives had separate burial areas, and non-denominational zones were allocated to white and colored (e.g. Cape Malay) populations (Christopher, 1995:40–41). In America, segregation could occur by denoting areas within the cemetery by race (Price, 1966:206). In some parts of America, however, there was explicitly no such racial limitation (Sloane, 1991:83). The material implications of segregation and integration would be a rich area of study.

Burial within cemeteries could be in perpetuity or could be only for a limited occupancy. In Britain and America, perpetuity was the norm, though in many British municipal cemeteries it was in fact 99 years (and thus is now becoming an important subject of debate in the matter of reuse or conservation of historic cemeteries). Additional payment could be made for perpetual care of the plot, which was particularly popular in America and Australia (Sloane, 1991:109; Zelinka, 1991:99). This can be very effective, and with appropriate investment can lead to an affluent cemetery as became the case of Mount Auburn, Boston (Linden-Ward, 1989a). In many cases, however, the funds obtained for burial plots and any continued maintenance have been spent up, with consequent decay and neglect as burial income declined or was insufficient to meet the costs of ever-increasing maintenance commitments. Some cemeteries have become bankrupt and been taken over by local authorities or trusts, particularly in Britain (Barker, 1984; Brooks, 1989; Murray, 1991); in America, many older cemeteries could no longer be maintained to the standards promised to those who had made perpetual care arrangements (Sloane, 1991:191–206).

In much of Continental Europe the solution to the burial crisis and overcrowded churchyards was the also the development of cemeteries, but there was an additional feature which ensured effective use of the space. A time limit was imposed on the length of time a body could occupy a space, often only about 30 years (Goody and Poppi, 1994). There was the opportunity, however, to pay an additional fee for perpetual occupancy, but this was rarely taken up in this period except by the most affluent who erected family mausolea.

Impermanent burials were of two main kinds in European cemeteries. Some were in dug earth graves, this being the overwhelmingly dominant form in northern Europe, and used for some burials in the Mediterranean. In southern Europe, however, another form of body location was popular. Wall vaults or *loculi* were constructed above ground, standing four to six storeys high and into which the coffin is inserted head or feet first (figure 8, left). After a period of years, which varied from place to place depending on the time taken for bodies to decay in the soil or in the *loculi* (Goody and Poppi, 1994), the space could be reused and a new body could be deposited. Any surviving bones from the previous interment were removed and placed in a communal ossuary, or just placed to the rear of *loculi* before another coffin was inserted. The wall vaults usually formed large ranges running around the walls of the cemetery, with earth dug graves and family vaults and mausolea in the open areas in the center, often divided by paths and softened by selective planting of trees. Thus the use of above-ground burial created a very different appearance and feeling to the cemetery.

In Mediterranean countries such as Italy and Spain, the more affluent families owned a private vault and mausoleum (Goody and Poppi, 1994). Normally six coffins could be placed lengthways on shelves and were sealed by large panels. Once the six were filled, further additions replaced the oldest occupants, whose remains were placed within a small family ossuary underneath the structure. Vaults may be partly below ground, or completely subterranean, but operated in the same way as the family mausolea. They occur widely and have been recorded in Gibraltar (Mytum, 1993). Mausolea were common in the New Orleans cemeteries (Florence, 1997; Masson, 1993), and more widely in Louisiana and Alabama (Sledge, 2002), but usually held two coffins at a time, separated one above the other by iron cross bars or stone slabs (figure 8, right). *Loculi* also occur widely in New Orleans cemeteries in communal blocks and society tombs (figure 8).

5. INTERNAL MEMORIALS

Many of the British 18th-century internal wall memorials were taken away in late Victorian restorations, and others have been moved from their original locations and placed elsewhere on the church walls. Some of these movements were with positive intent, such as away from behind a newly installed organ, but others were placed out of sight such as in the church tower, either to remove what was perceived as clutter in the church, or to make room for new memorials. The same fate applied to many floor ledger slabs, either as new burials took place or as the slabs became worn. The installation of heating systems, new blocks of pew seating, or the laying of decorated tile pavements also spelt the end of floor

Figure 8. Left, Block of *loculi* where coffins can be placed end-on into chambers, Granada, Spain. Right, Family mausoleum divided so that coffins can be placed one above the other, La Fayette cemetery, New Orleans, Louisiana.

ledgers in their original settings. Thus, many studies of these monuments rely on contemporary documents and subsequent antiquarian descriptions and illustrations of churches as well as field evidence.

5.1 Materials

Internal wall memorials became popular for those buried in vaults and crypts, and white marble was the overwhelmingly dominant material until the late 19th century when, partly through Gothic revival influence, brass became a popular alternative (Meara, 1983; 1991). An artificial stone, Coade Stone, was used for some elaborate internal and a few external memorials in the late 18th and early 19th century (Kelly, 1990:243–256).

Investment in commemorative church furnishings and stained glass increased rapidly during this period (Rawding, 1990). These forms of memorial are not considered further here, but were an important opportunity for competitive display, patronage of arts and crafts, and in the elaboration of places of worship.

They were all were subject to changing fashions and liturgy, and so many items have been moved or replaced over the decades (Harrison, 1980; Rawding, 1990). Not directly linked to the burial, there was less pressure to preserve these items. Nevertheless, such commemorative patronage was extremely important in revitalizing church ornamentation following first the Reformation, then the English Civil War, and finally the widespread lack of investment in church buildings during the 17th and 18th centuries. The great wave of Victorian church restoration and rebuilding, encouraged by the Gothic Revival, provided both a spur and a context for commemorative stained and painted glass on a scale not matched since (Harrison, 1980; Sewter, 1975).

5.2 Form and Style

The numbers of internal memorials rose significantly during this period in both Britain and Europe, though most were not large or elaborate. Ledger monuments were by far the most numerous internal memorials, placed over the brick-lined shaft or entrance to the family vault. As internal burial was discouraged and then prohibited (the timing varying from country to country), so the popularity of the ledger waned during the late 18th and 19th centuries. By then, many urban church floors were completely covered with inscribed ledgers, and they were common also in rural areas. Whilst some had carved decorative designs or inset brasses, most were simply incised with text.

Wall monuments in all styles were stock items produced by sculptors in the main towns and cities, though there was still a flourishing market for specifically commissioned pieces containing portraits and more elaborate carved scenes (Whinney, 1988). Some 18th-century memorials were highly complex and major works of art, and as such have been subject to considerable attention by art historians (Llewellyn, 2000; Whinney, 1988). Such monuments, with some local differences in elite taste, are found throughout Europe but do not frequently occur in North America. Here relative poverty, lack of suitably skilled artists and craftsmen in this milieu, and differing attitudes to commemoration prevented the development of such products. By the 19th century, large internal memorials were rare, as the cemetery allowed monumentality in an external setting, and public statuary and other forms of structures were erected in non-funerary contexts (Bruce,1997). Whilst it was the British and European 18th-century internal memorials that were the most extravagant of their time, it is American 19th-century cemeteries that provide the most opulent external monuments (see below).

In Britain, the role of Westminster Abbey as a burial and commemorative location for the heroes of the Empire became established, with numerous iconographically rich monuments mixed with others of great simplicity, though all of the highest quality (Whinney, 1988). In Europe some other religious buildings such as the Vatican and cathedrals in capital cities have large concentrations of such monuments.

Figure 9. Two internal wall monuments with texts on panels shaped like sarcophagi, with urns above. Note the presence of some heraldry. Left, Salton, North Yorkshire, England. Right, Termonfeckin, Co. Louth, Ireland.

The dominant style for internal memorials throughout this period was neo-Classical (figure 9), sometimes greatly elaborated but in many cases only indicated by the sarcophagus shape of the panel or the presence of a mourning figure in Classical garb. Urns were also very common, as were cherubs, though skulls still occurred occasionally. Heraldry also is present, though is rarely dominant; it did not sit so easily with a strong Classical theme. In many parts of Continental Europe, Baroque and Rococo were the dominant styles, though many of the same motifs and symbols were used whatever style was employed.

Gothic revival internal memorials also became popular during the later 19th century, with revival style chest tombs at the more elaborate end of the range. Wall monuments tended to merely utilize gothic motifs and lettering as there were no medieval precedents. These were usually in stone, incorporating tracery and ball flower designs not dissimilar to some found on external monuments including headstones. The use of brass was inspired by Pugin but followed by others (Meara, 1983; 1991; Whittemore, 2001). A small number of internal monuments followed other revival styles such as Egyptian and Byzantine, but these were more popular outside. Some memorials were eclectic mixtures of style, a feature notable in the architecture of the time and seen on some external monuments.

Wall memorials of all these types occurred widely across Britain and Europe and in the main colonial churches and cathedrals. Relatively little has been published on these, but some note has been made of those on the Indian subcontinent (Muthiah, 1990:11, 13), New Zealand (Fields and Stackpoole, 1973 passim), and the Caribbean and Belize (Coutu, 1997; Hammond, 1999; 2000). They were rare but not completely absent in America (Dorsey, 1957:25). By the 19th century, many internal monuments were complemented by external memorials that marked the actual site of burial. In these cases, the presence of a commemorative feature in two arenas allowed for different forms of social presentation and remembrance.

6. EXTERNAL MEMORIALS

It is during this period that the external monument becomes so ubiquitous. In the later 18th century, only a small proportion of the population were commemorated in a permanent way, but by the late 19th and early 20th century the burials of well over half the population in many areas were so marked. Given the rich material evidence for this period, it is one that has attracted considerable attention by graveyard recorders, though there is proportionately less published than for the earlier period. This is in part due to the increased use of pattern books and standardization which has meant that many with aesthetic concerns have found the memorials less rewarding. There are, however, still many regional variations in monument choice, treatment, and materials used. Moreover, the greater social range of those represented by memorials allows for more inclusive studies and for examination of minorities not previously represented.

Monuments away from burials became more common and prominent. Memorial figure sculpture became a major feature of the townscape during the 19th century, with figures in stone or bronze adorning many public places (Read, 1982). Thus the great and the good could be remembered by the general populace, but not at their actual grave sites. Wealthy individuals often preferred to bequest endowments enabling new construction or rebuilding, including places of worship, education, and learning. This phenomenon can be seen not only in Britain but across the colonies and America, where it became particularly popular and was often achieved during the benefactor's lifetime. Commemoration thus could take many forms, but monuments at burial locations were still the most frequent material manifestation.

6.1 Materials

During the 18th and early 19th centuries, local stone was used if it was suitable for memorials. Gradually, however, a wider range of materials became available (figure 10). At first marble and then granite came to be used on more affluent

Figure 10. Four headstones of different materials. From right to left, yellow sandstone, pink granite, gray granite, white marble. A flower vase has been placed in front of one headstone. York cemetery, North Yorkshire, England.

monuments, sometimes as part of a polychrome effect. Fine-grained sources such as slate also began to be selected more frequently beyond the areas immediately around the quarries. During the 19th century certain rock types became favored for memorials, and national and even international trade developed. In Britain, Scottish granites and Italian marble became very fashionable, eventually eclipsing local sources (Kennett, 1995; Robinson, 1994; 2001). Even in areas such as Pembrokeshire where fine slate was available, this trend has been identified (Mytum, 2002e). As colonial sources were discovered, more distant materials were exploited, though this trend and the use of Italian marble became most noticeable after World War I. Polychrome headstones also appear during the 19th century, with insets near the top or on the sides of the stone. These are most frequently marble, but other rocks of contrasting colors, or even glazed tiles or terracotta insets, can be used. Polished igneous rocks were popular for engaged columns on headstones and tombs, a feature also found on contemporary buildings.

The same pattern of increased range of rock types has been noted in North America. This was first exemplified by the introduction of marble to augment sandstone and slate. Gradually, however, granites and marbles became the dominant materials, and masons were accustomed to work in a range of materials. The trends in Ontario have been summarized by Norris (1988:135), where limestone was the overwhelmingly dominant material in the 19th century. The minority materials were slate in the first half of the century when they made up about 10% of the stones, and sandstone which had 30% of the market in the first half of the century and 20% thereafter. Granite first appeared around 1870, and rapidly expanded to 25% by 1890, and was used for over half of the memorials by the early 20th century (Norris, 1988:131, 135).

The affixing of bronze, brass and other metal alloy decorative features onto stone monuments was common in Europe and to a lesser extent elsewhere. This could also include inlaid or raised lettering of the inscription. Decorative features may include portrait busts and wreaths as well as religious symbols and iconography. This tradition of memorials was often maintained by emigrant groups when they moved to colonies, Britain, or North America.

6.1.1 Wood

Wood was still used for memorials to the poor, and in those areas where the funerary industry had not become established, but usually only in the form of a simple cross. The use of evergreen trees within and adjacent to plots increased, however, and in some places these survive even where the memorials have disappeared. The post and rail monuments continued, but other wooden monuments also occurred. Crosses in wood were popular; some were left blank, but others had a carved or painted text. By definition, the survival of impermanent markers is rare, but it would seem that some headboards having the same function and appearance as small headstones were erected. In all but arid environments such as the American southwest, survivals are rare, and even there the erosion of the surface often means that an early date has to be assumed rather than proven (Jordan, 1982; Little, 1998). However, a few examples have been noted in South Carolina (Rauschenberg, 1990), and an as yet unpublished example has been recovered from a waterlogged context in Ireland. In some countries such as Hungary, wooden memorials were a traditional form of memorial, particularly where suitable stone was scarce, and these could be quite elaborate. Muslim memorials could also be of the pillar form, but surviving examples are in stone. The Himba of southern Africa incorporated stones, ox skulls and wooden poles, carved and tied together (Bollig, 1997).

6.1.2 Metal

Manufactured goods in large quantities and at relatively low prices were a result of the industrial revolution. The potential market in memorials was seized by a range of manufacturers who created monuments that either mirrored those available in stone, or used the advantages of the new materials to create different

Figure 11. Left, Cast iron memorial onto which commemorative text would have been painted. The metal legs fixed it into the ground. Sterling, Scotland. Right, Cast iron cross with applied elements including Crucifixion and roundel for commemorative text. Note the use of paint to highlight the figures. Rural graveyard, Czech Republic.

forms. For example, cast iron memorials were occasionally produced in ironworking areas of England for interior locations from the 16th century, and outside from the early 18th century (Willatts, 1987). But it was only in the 19th century that it became a more widespread material, with single-cast items, or pre-fabricated sets of elements to make up more complex pedestal monuments (O'Neill, 1965). Many of these imitated stone monuments, though usually with raised letters as these were impressed into the sand from which the monument was cast. In Britain, small but often highly decorated cast iron memorials were produced as a cheap alternative to headstones (figure11, left). These could have the lettering incorporated at production, but many of the least expensive were cast in large numbers without text, which could be painted on at the point of sale. Unfortunately, corrosion and weathering has often meant that such monuments are now not datable. They are also small, fixed into the ground by thin bands which can corrode or be broken off, and many are now lost or out of position. Cast and wrought iron was also important for elements on complex memorials, and for memorial crosses in some countries (figure 11, right). Moreover, the large number of chains and railings around burial plots created a considerable demand for this material in the cemetery context (Forbes, 1990; Gayle, 1990; Linden-Ward, 1990).

Zinc, termed in the contemporary literature as white bronze, was popular in cemeteries in Britain and North America from the 1870s, though its use declined after World War I (Rotundo, 1989). The Monumental Bronze Company of Bridgport, Connecticut and its products have been studied, and they were largely imitations of stone products, but could include decorative elements that would have been impossible in any stone. A wide range of items were produced, allowing many social classes to have this material within their range of choices, but it never became really popular despite its competitive price.

6.1.3 Ceramics

Ceramics were also used for a few memorials, but usually the material was incorporated into stone as applied or inset ceramic tiles, or terracotta panels (Remnant, 1962; 1964). Some monuments were made completely of terracotta as an inexpensive alternative to stone. Always a minority, such production centers were usually short-lived and have attracted little attention. One example that has been studied is that in New Jersey and New York where a wide range of forms, often highly decorated, can be found from the late 19th and early 20th century (Veit, 1995). Concrete in the 19th century tended to be restricted to forming the base of monuments, or for making the kerbs that define the plots. These materials became more popular in the 20th century. African-American graves often incorporate ceramic and concrete elements, but relatively few extant memorials of this type are certainly as early as this period. There are sufficient survivals to suggest, however, that the traditions of home-made memorials using materials such as concrete were established during the 19th century, combining traits a range of diverse cultural sources (Little, 1998; Vlach, 1978:145).

6.2 Forms

Just as the traditional repertoire of materials continued for a considerable period, the same was the case with forms. Indeed, the early bedstead and shouldered semi-circular headstone designs survived as a minority throughout the period. Some general trends in the popularity of forms over time can be outlined, but there were clear regional and national differences. The greatest emphasis of research to date has been focused on the origins and development of particular regional forms. These have often been placed in a spatial context and related to craft production, ethnicity and other forms of identity; these studies are reviewed in the thematic chapters of the book. Relatively few typological studies of all monuments in a region have been published, but studies from Ontario, Florida and Britain provide some indications of the sorts of trends that can be expected.

The headstone form remains the most popular in many areas. In Ontario, the obelisk is the major alternative (Norris, 1988), but in north Pembrokeshire it was the cross which provided the most important larger vertical form, though tall monuments including columns and obelisks were visually notable if not numerous (Mytum, 2002e). The chest tomb was also more popular in some areas, but not in Ontario. North American assemblages show the importance of the scroll monument and tablets (Dethlefsen, 1981), which continue in popularity well into the 20th century. Scrolls are also found in Britain in a range of sizes and styles (Mytum, 2000:113), though rarely identical with those so commonly found in America. Small tablet headstones that were so common on family plots in America do not occur in Britain, though a few diminutive headstones were selected, mainly for children.

Due to the cultural expectations for greater standardization and the influence of pattern books, however, many monuments of very similar form were produced over vast areas at the same time. Although locally made, little often distinguished individual monuments from those elsewhere on the globe. Thus, a headstone in a cemetery in Sydney, Australia could be identical to one in Boston, America or London, England. Moreover, as certain quarries began to specialize in monuments, the initial shaping would seem to be carried out at the quarry sites, and carved blanks distributed. However, little documentation survives that specifically deals with the funerary market. Graveyard and cemetery monuments were generally a regular but small part of the architectural business of a quarry. Thus, Cornish granite is noted for its use in Nelson's column, London, but the numerous monuments also using this rock have not received attention (Stanier, 1985).

6.2.1 External Mural Monuments

In the 18th century, some monuments were placed on the exterior walls of churches in Britain and Ireland. Such monuments also seem to occur on Continental churches. These memorials often combine some characteristics of internal memorials and those of external headstones. A limited number of external memorials were also erected in the 19th century, but were not common. All mural monuments reflect a degree of social status greater than the overwhelming majority of contemporary external memorials. The well established mural traditions in Scotland continued with some substantial monuments, but small wall tablets also were erected, and some headstones were placed against boundary walls mirroring this style (Willsher 1985b).

6.2.2 Post and Rail

This form was continued in the same areas of south-eastern England where it been popular in the earlier period. In the 18th century the post and rail memorials were often made in stone, with a range of chunky forms necessary for this material

Figure 12. Late example of a wooden post and rail memorial, Moulsford, Berkshire, England.

(Burgess 1963). During the 19th century examples also occur in cast iron, with the lettering on the board formed by pressing type into the sand mould for that element of the monument prior to casting, producing a similar effect to the raised lettering on the 1658 wooden example. The cast iron examples also returned to thinner, more elegant proportions because of the material used. The form seems to have gone out of use in the early 20th century (figure 12).

6.2.3 Raised Platform

Nigh (1997:167) has published an illustration first published in 1887 showing a platform raised up on timbers on which a shroud-wrapped body was exposed. It is suggested that this reflects African traditions within an African-American mortuary tradition. At one end of the structure an animal skin is draped over a post, and scattered around it are various vessels. which have been broken or punctured. Various features of this scene suggest continuity of traditions with regions of Kongo and Angola (Nigh, 1997:165–168). Moreover, this type of body disposal could explain the relative infrequency of slave burials at some plantation sites, if such practices were allowed. The increasing influence of Christianity and the formal controls over burial areas means that such traditions do not appear to have been widespread in the 19th century, and do not survive into the 20th century. Nevertheless, aspects of ritual and cultural behavior do continue. Four types of African-American grave marking have been noted by Little (1989) in North Carolina, and these are largely from the 20th century and so are discussed in the next chapter. She considers, however, that all were probably in use in the 19th century, and so early evidence concerning the use of these types in this period would be very valuable.

6.2.4 Headstones

During this period, headstones are the dominant monument type in many regions, though over time other forms do become popular. Moreover, the headstone changes significantly during the period. By the later 18th century architectural traditions were more dominant in influencing form, and has been noted most frequently for British memorials (Burgess, 1963; Mytum, 2000). Moldings were either cut to give a three-dimensional quality to the memorial or, in stones such as slate less able to withstand this type of treatment, designs were incised to give the impression of architectural features.

From the mid 18th to the early 19th century, regional styles can still be discerned in the design and production of headstones. Whilst the folk style carvings become less common during the later 18th century, variability is still considerable, though the placing of designs on both sides of headstones becomes extremely rare in this period. Studies of New England carvers of this period have been extensive (e.g. Benes, 1977; Chase and Gabel, 1997; Ludwig 1966; Slater, 1996a; Tashjian and Tashjian, 1974), though their different products are distinguished within a broad trend in changes of motif discussed below. For example, the various 'bedstead' forms give way to rectangular or simple shaped stones, linked to a neo-Classical tradition of the *stela*, and so mirroring the aesthetics seen on internal memorials. Other carvers have also been examined, including some in Canada (Trask, 1978b).

During the 19th century, revival styles of architecture largely dictated headstone shape, a trend which continued into the early 20th century. Neo-Classical forms retained their popularity, but were usually outnumbered by Gothic stones in the latter half of the 19th century. The advantage of this style was that it could be combined with a wide range of decorative features in the same style, though the combining of different styles in form or decoration was far from unusual. The Egyptian revival tended to be reflected in decorative motifs and different monument forms such as obelisks and mausolea rather than in headstone forms, but the Romanesque and Byzantine revivals also influenced headstone design in some regions.

In North America, more alternatives to the large and relatively thin headstone were developed. The small, round-topped tablet form became extremely common, usually carved in marble. This was the form used in the National Cemeteries, but

Figure 13. Tablet headstones in a military cemetery, St. Augustine, Florida. Note pyramid communal memorials to the rear.

it was used widely in other contexts (figure 13). Other small forms were the scroll, similar in size to the round-topped tablet but largely of an open scroll form, though its angle varied greatly, and only the more vertical forms could be considered in the headstone tradition. Other forms, less related to headstone because of their thickness, are discussed under low monuments. They form part of a distinctive late 19th and early 20th century North American cemetery repertoire with some parallels elsewhere, but diagnostic largely because of the style of family plot which became dominant in American cemeteries.

Headstone shapes are extremely diverse, and the ways in which such variability should be recorded and analyzed are also numerous (Mytum, 2000). Whilst the styles described above form one strand in the design repertoire, other localized factors influenced forms. The hand-made nature of headstones within a mass market means that they can be examined on a scale that ranges from unique items to very broad categories. This is true also of some but not all other categories of monument, but is far more obvious with headstones.

6.2.5 Crosses

The stone cross became a significant minority selection in the late 19th century in Britain, its colonies, and North America, and grew in popularity during the first part of the 20th century. The crosses occur in a wide variety of forms. One common type, particularly in Britain, was a design with a stepped base, which grew in popularity during the first part of the 20th century, and could be relatively small (Mytum, 2000:108). Celtic ringed crosses were chosen by Irish at home and abroad, as part of a late 19th-century nationalist Celtic revival, but were also chosen by others without any overt political or cultural motivation. Other revival crosses include Gothic floreate crosses, but other types include Anglian and Scandinavian designs. In Continental Europe, and amongst emigrée communities elsewhere, other cross forms were used extensively in wood, metal or stone.

Metal crosses occur in many burial grounds during the 19th century. They may be substantial cast iron crosses found in Victorian cemeteries, or much cheaper memorials which are smaller and cast in iron or other alloys. Crosses were more popular on Continental Europe, where in some regions they were the standard memorial. They are widely found in Catholic France and Germany (Brown, 1994:217; Kselman, 1993). Metal crosses are also common in the Baltic states, and occur elsewhere in Europe. There is a strong German tradition for wrought iron markers, such as those from Bavaria, with the cross taking a number of forms and elaborated with other decorative elements and symbols such as hearts (Brown, 1994:208–218; Patterson, 1976; Wright, 1980). Some crosses have the coping which is often seen in wooden crosses in Europe and on emigrée burials. Cast iron crosses also occur (figure 11, right), some with the Crucifixion (Horton, 1997). Such crosses and Crucifixions also occur fixed on wooden or stone monuments.

Wooden crosses were widespread. In Continental Europe where earth burial was the norm, the wooden cross was the item of choice for those able to expend some resources on commemoration at the grave but were not in the minority who could afford a more monumental stone or metal feature. Most of the memorials erected in Père Lachaise, for example, were simple wooden crosses (Kselman, 1993:184), though cast and wrought iron crosses became more popular throughout France as the costs of these dropped with industrialization. Crosses could often be more complex and decorated at little additional cost; many were also painted and it is unclear whether wooden crosses were also so treated. Wooden crosses were also the popular form of marker in New Mexico *camposantos*, though often difficult to date. Warren (1987:116) notes that from the 1880s commercially available timber and paints made it possible for more complex and colorful designs and structures to be created, and wooden grave fences were made as an alternative to the expensive and therefore rarely used iron railings. These traditions, even in this relatively poor and isolated area, reflects the same desires to mark graves (relatively permanently even in wood in this arid zone) seen elsewhere. In Texas, a wide array of cross designs has been noted by Jordan (1982). Metal crosses were erected mainly in the German graveyards, and wooden ones in the Mexican burial grounds.

Figure 14. Left, Two tombs, on the left with an urn, on the right an obelisk. Note the white marble bust on the pedestal base of this monument. Bradford Underfcliffe cemetery, West Yorkshire, England. Right, chest tomb with sarcophegus on top, Msida Bastion cemetery, Malta.

6.2.6 Pedestal Monuments

Pedestal monuments appear as part of the Baroque and Classical revival repertoires, in a variety of 18th-century forms. Many pedestal monuments are box-like structures which are more vertical in form than chest and table tombs, and do not have a top slab on which inscriptions may be placed. Some may have simple coped tops, but many had additional features such as finials and urns. During the 19th century the range of common additions included columns and obelisks (figure 14, left), the latter particularly popular in North America. Extremely common on the Indian subcontinent, pedestal monuments were often made of local stone with marble or other stone inset on the sides for inscriptions. Some of these were of great size, and merge into those categorized as tombs. These monuments can be found in all types of British burial grounds on the sub-continent, from urban sites such as Kacheri cemetery, Kawnpur (Yalland, 1985) to hill station graveyards such as that at St. Stephen's church, Ootacamund (Davies, 1986:123–124).

Pedestal monuments occur in rural graveyards but were most common in urban cemeteries where competition for visibility across burial areas packed with memorials was greatest. Many monuments of this type were made in specialist workshops and bought in by local masons who added the inscriptions. Most pedestal monuments were thus constructed of non-local materials and were often made with materials from a number of different sources. The range of colors of the sandstones, marbles and granites used for such monuments further helped to distinguish and identify them in the cemetery context, though where many were placed close together this individual distinctiveness could be lost.

6.2.7 Tombs

Tombs became increasingly popular during the 18th century and remained so through much of the 19th century, though they became plainer as time passed. Table tombs continue as a minority choice, though examples with exotic granites or cast iron legs reflect the increased range of materials used in memorials. Chest tombs continued to be the dominant form, and could be in a design that reflected a range of revival styles. Some tombs were modeled on Classical sarcophagi (figure 14, right), but others were inspired by the medieval chest tombs. Most chest tombs were relatively plain, with two main traditions. The simplest relied on the tabletop stone being used for the commemorative inscription, the more complex had panels in the same stone or insets of a different material on the sides and ends of the tomb, and could carry text.

Tombs were always more frequent in Britain than in North America or Europe, and this is probably due to the contexts of burial. Tombs of table or chest type tended to be the most flamboyant features of many churchyards, and they were the greatest level of monumental investment seen as acceptable. Grander monuments such as large figural sculpture and mausolea were encouraged in cemeteries, and so the variety and extravagance of monuments was greater there, and few desired the chest tomb in such contexts. Indeed, this can be seen in North American east coast where the older graveyards contain a scatter of chest tombs, often with brick or ashlar masonry base and simple slab with inscription on top, in a similar manner to those in British churchyards. In the later cemeteries in this area the chest tombs are relatively rare, as they are in the British cemeteries.

In parts of the American south such as Florida, Louisiana, Alabama and the Carolinas, a range of brick tombs developed that in their larger forms became mausolea (see below). These tombs tops in a range of shapes (figure 15), and the inscribed slab, usually of marble, was often placed vertically at one end of the tomb, though usually this does not survive. This has inhibited the development of detailed chronologies, but some valuable identification and classification has been undertaken (Florence, 1997; Little, 1998; Nakagawa, 1994; Sledge, 2002; Thompson, 1993). In many respects these monuments in brick mirrored the partially buried tombs of the Mediterranean, though these were usually clad in marble and so have a more finished appearance. Similar traditions also seem to have been prevalent within colonial communities in South America (Zucchi, 1997).

Figure 15. Brick tomb, Church Street graveyard, Mobile, Alabama.

Figure 16. Partly subterranean tomb with marble slab over the access point. This particular tomb is used for those who do not have their own family tombs; the plaques on top commemorate the various individuals within the tomb. Pollença, Mallorca, Spain.

6.2.8 Ledgers

Ledgers continue to be used by a small minority, and sometimes sit on a low base. Smaller, less elongated flat slabs also occur both in Europe and North America, though are rare at this period in Britain. These may be of more than one material, with an inset panel for the inscription. They are often for single interments, whereas the size of ledgers gave the opportunity for more commemorations.

6.2.9 Low Monuments

Although there was a general trend during this period towards greater height and visibility for memorials, a range of low monuments became extremely popular as an alternative to the tomb or ledger (Mytum, 2000:102). Many were in a revival style, and indeed the inspiration for the form came from the antiquarian excavations at abbey sites where large numbers of medieval grave slabs were uncovered. These designs became a popular Gothic revival form, usually with a floreate cross and often with Gothic style lettering; they tend to be in sandstone, limestone or marble. Equivalent low monuments were also produced in a neo-Classical style, but the most common other low forms were simple coped granite examples. In America, these monuments were relatively rare, but low, squat stones incorporating a scroll, open book or beveled face for inscriptions became popular on the family plots within cemeteries.

6.2.10 Additional Elements

During the 18th century monuments often consisted of several elements, and this tendency increased during the 19th century (figure 17). The most common additions were footstones, which had a long tradition, but were often more elaborate at this period. They may carry the initials of the deceased and often the year of death. Sometimes, but by no means always, they mirrored in miniature the shape of the headstone, and a few had decoration such as Gothic tracery. Body stones or ledgers that covered the grave were also common, and could be used in conjunction with footstones. Some flat stones were decorated, some made in the shape of coffins. In Europe, many graves had a ledger raised on a low base which sloped down from the headstone and provided an easily visible surface on which the names and dates of family members could be recorded (Stokroos, 1992, passim).

In the 19th century, the footstones and slabs continued, but kerbs became a common additional feature defining the plot (Mytum 2000:118). In many cases these kerbs were sufficient, but elaboration in the form of iron railings or posts and chains was popular. Such features could be found round all monuments, even mausolea. These features can be seen throughout the world, but their survival varies depending on later graveyard maintenance and security. Many have been

Figure 17. Headstones with footstones and body stones, all within a low walled plot. Church Street graveyard, Mobile, Alabama.

stolen or removed to ease grounds maintenance, though their rarity in Britain was caused by large scale clearance as part of the war effort in World War II. The most elaborate headstone graves could have all these additional features, making them as expensive and dominant as other forms of larger monument for which no additional features were supplied.

6.2.11 Mausolea

Some mausolea were constructed in churchyards, but most were erected in cemeteries where they could be shown off to good effect and were often grouped in elite zones. Throughout the industrialized world, the newly affluent upper middle classes solidified their dynastic success in a family mausoleum (Fenza, 1989; McDowell and Meyer, 1994; Wright, 1988). These mausolea were of two main functional types, those with the bodies above ground (a largely Mediterranean tradition) and those that had a below-ground crypt beneath (northern European elite burial tradition). From the outside these two types cannot always be separated, as the style of the monument could be identical.

Revival styles dominated the market for mausolea (Broman, 2001; McDowell and Meyer, 1994). Three styles were most frequent, with Neo-Classical temples (often marble or limestone), Gothic revival chapels (often sandstone) and

Figure 18. Society tomb for firemen, decorated with a fire engine. La Fayette cemetery, New Oreans, Louisiana.

Egyptian structures (almost always granite). Most British and European examples were of modest scale, though in North America some were of great size (Brown 1994; Florence 1997; Linden-Ward 1989a). Eclectic and unusual mausolea occur widely, however, as those with sufficient funds to afford such a monument could chose to indulge their taste (or often lack of it) in an extravagance which ensured that their memorial would attract attention. Some cemeteries had strict controls on monuments, but most welcomed the income that the sale of the plot and permission to build could bring.

6.2.12 Loculi

In much of Mediterranean Europe and also Louisiana and Alabama, the practice of placing coffins end-on within upstanding small chambers became frequent. As a response to the burial crisis they were very effective. Though they involved a greater initial investment than earth burial, they had the virtue of easy maintenance and control. Moreover, they operated on a limited tenure basis so that the remains could be pushed to the rear or removed to a communal ossuary and the chamber reused after a period. *Loculi* were built in blocks and so had a strong architectural presence, though some were also placed underground as catacombs which remain open to public view.

Most *loculi* were communal and run by the cemetery management, but some, particularly in America, were constructed by groups, often based around employment (figure 18) or religion, and are often termed society monuments (Florence, 1997). These monuments could be extremely elaborate, and there was clearly pride and competition associated with such structures. The openings of the vaults were bricked up, and could be plastered over or have a memorial tablet placed on them, giving some personal details. Such tablets were not retained, however, when the vault was reused, and so the early memorial styles are largely unknown. The coverings on the society monuments tend to be uniform, maintaining the group identity, but those on the communal blocks can vary greatly in material, decoration and use of color. No systematic research has been undertaken on the stylistic changes of such memorial tablets, but observation in mainland Spain, Majorca and Italy indicates that there were changes over time.

Loculi are sometimes called "oven vaults" in New Orleans, where this style of interment seems to have been copied from Europe (Florence,1997:24–26); though it is claimed the high water table may have been a major factor this is more likely a post-hoc explanation of a form otherwise not widely found in America. The widespread use of *loculi* in the region on sites not all affected by water table suggest cultural factors were the dominant cause; water table problems were not common in the Mediterranean.

6.2.13 War Memorials

Memorials have been erected for the war dead after many conflicts, though a large number were not constructed until a considerable period after the event; this is particularly the case, for example, with Revolutionary war battles in America, and Washington's tomb at Mount Vernon (McDowell and Meyer, 1994). The American Civil War led to the establishment of cemeteries for the dead from the conflict, these being the first specialized yet public burial grounds of this type, with distinctive forms of communal memorial as well as markers for individual interments (figure 13). There were also separate memorials to the war dead (Mayo, 1988). Elsewhere, war memorials were not a feature for several decades and were rare in Britain before the Boer War.

6.2.14 Impermanent Materials

The most obvious form of grave marking which has no inscription is the mounded earth over the grave. Though this may sink to a certain extent as the coffin collapses and the body decays, it often survives as a ridge in the ground. This mound may be enhanced with rocks or shells, and these may be arranged around the edge of the mound or in a design on the top. In areas with good grass growth, the turf cut prior to digging the grave may be used to consolidate the mound, and can create a notable and lasting feature. Unfortunately, such markers have no date, and so their use in this period can only be confirmed from contemporary illustrations or descriptions.

Vegetation can be used to mark graves, rather than enhance a more complex memorial. Some cultures commonly mark graves with particular plants, and the use of cacti or shrubs in some slave burial grounds is an example of this. Whether this reflects an African tradition, or that this was a local response utilizing options available within the American context in uncertain.

Uninscribed stones used as head or foot stones can still be noted in many graveyards, and their use continues to the present day. They were clearly popular during the 19th century as they can often be found in burial grounds only set up at that date, and can be seen interspersed between inscribed memorials. In Ireland separate burial grounds for unbaptized infants, *killeen*, are normally indicated merely by a scatter of small protruding stones, often placed within an ancient enclosed site such as a church or monastery. The origin of such sites is uncertain but they continued in use until the middle of the 20th century (O'Brien and Sweetman, 1997:121).

7. DECORATION AND SYMBOLS

During the 18th century, the two dominant and widely found decorative features are the cherub and the urn. The researches of Deetz and Dethlefsen (1965; 1967; 1971) have established the chronological and spatial changes from death's heads to cherubs and then to urns and willows in New England (figure 19). The clarity and chronology of these shifts are not found everywhere, however, and mortality symbols last through much of the 18th century as a minor element in the repertoire in many regions, and indeed occur rarely even later. Also, within both the dominant symbols of cherubs and urns there is an almost infinite variety of representation and emphasis. Many studies have attempted to use the treatment of decoration and symbols in this period to identify particular carvers, and this aspect is discussed in Chapter 5.

The transition from death's head to cherub has been investigated most minutely in New England, where the links to changes in Puritan thought have been seen as a force for change (Benes, 1977; Ludwig, 1966), possibly linked to wider shifts in world view suggested by Deetz (1977). This change is less visible in Britain and Ireland where only a few regions have large numbers of death's head stones (Burgess, 1963; Mytum, 2000), and where the overlap and combination of cherubs and skulls lasted into the 19th century. The meanings implied by the use of these symbols in non-Puritan contexts were undoubtedly different from those discussed for New England. Nevertheless, many of the design elements had an extensive distribution, as sources of inspiration such as book engravings had a wide circulation.

Figure 19. Left, Headstone with mortality symbols, Boston, Massachusetts. Right, Headstone with urn and willows motifs, Ipswich, Massachusetts.

Cherubs may be represented with or without wings, and may be portraits or whole figures, and can be plump *putti* figures of the Baroque, or slimmer and more agile mature figures (figures 34, 45). They may be passive, expressing a range of emotions, or can be active, undertaking a wide range of tasks. These include holding shrouds, warning texts or symbols of mortality such as hourglasses. They can be representing the soul leaving the body of the deceased, or can be accompanying the Resurrection, blowing trumpets and helping the saved to their eternal rest. By the 19th century it can be difficult to separate some depictions of cherubs from those of angels.

Urns are the other major decorative motif of the 18th century, and like cherubs come in a bewildering array of forms. Some were simple and tall, others squat and dripping with foliage and drapes (Burgess, 1963:175). In America and some parts of Britain such as North Wales, the urn was usually accompanied by a weeping willow, but elsewhere this was not the case. The urn was most frequently used alone, or with other features such as a weeping female figure, who could also be illustrated with an altar instead of an urn (figure 20). Urns formed part of a developing Neo-Classical iconography associated with mortuary monuments, and in the later 18th and through much of the 19th century became the most

Figure 20. Headstone with mourning woman beneath weeping willow, Holyhead, Anglesey, Wales.

popular but by far from dominant motif. Other common design elements included flaming torches, serpents eating their tails, shells, and many Classical architectural decorative features such as the palmette, acanthus, key patterns and other elements of frieze designs.

The Gothic revival in the 19th century led to a large increase in the repertoire of motifs. Most were based around the cross or architectural features such as tracery. The IHS symbol, which had been carved in relatively simple forms on Catholic monuments, also came into vogue in more elaborate Gothic lettering, often as a monogram. The new motifs could be treated in many different ways, in simple incised designs or deeply carved three-dimensional renderings; they could be simple allusions to a style, or forcefully conveying the theme. Elements of more than one revival style could occur on the same monument.

Other revival styles also had their own repertoire of decoration, such as dog tooth carving for the Romanesque, but the Egyptian revival monuments tended to be relatively plain with only the most elaborate mausolea and a few minor monuments having sphinxes and winged globes. Egyptian decorative features were popular at 19th-century cemetery gates, however, where pylons, columns and the cavetto cornice were frequently used (McDowell and Meyer, 1994).

In the 18th century a number of New England stones have portraits which clearly represent the deceased (figure 42), but such representations on memorials of that date are rare. Portraits, and indeed full size figural sculpture, becomes more common in the 19th century, and the implications of these are considered elsewhere. Scenes with generic figures found on more elaborate 18th and 19th-century stones are often linked to Classical or Biblical themes. Figures in contemporary dress are rare, though some of the Last Judgement scenes show settlements, churches and burial grounds as they were when the monuments were carved (Burgess, 1963). In Catholic contexts, Crucifixion scenes are common (figure 34) and in Ireland the symbols of the Passion were popular in some regions during the 18th and early 19th centuries (de hÓir, 1997). A variety of Biblical scenes have been identified, including the Nativity (Longfield, 1947) and Adam and Eve (Willsher, 1992a); symbols with more complex Christian and other associations may be more difficult to identify and interpret, such as the Green Man on Scottish memorials (Willsher, 1992b). The other major form of figural decoration on monuments was the mourner, usually female and dressed in Classical garb, though she could occur with elements from any tradition (figure 20). She is often draped around an urn, sarcophagus, cross or anchor, the last being a symbol of hope. More widespread are generalized, vaguely Classical mourning female figures (figure 21, left), sometimes as parts of elaborate monuments.

In 19th-century cemetery contexts some notable and successful figures have busts in terracotta or bronze affixed to their monuments. These are usually in low

Figure 21. Left, Mourning figure, Magnolia Cemetery, Mobile, Alabama. Right, Marble angel with cross, Brompton cemetery, London, England.

relief, but in some cases can be in full three dimensions. These busts and portraits also can be found on Continental Europe, and those of notable cultural and political figures still survive to this day, even where most burial plots have been successively reused and so their older memorials have been cleared away. In America, the use of busts is also frequent, but again only on the more affluent memorials. They would seem to be rare in Australia, though marble examples were placed on the memorial to Amy and James Barnet (Weston, 1989:117).

At certain periods, representation of the deceased is a favored feature on memorials. There is a Continental European tradition for the use of photographs on graves. This began with the early daguerrotypes, and has developed with the use of color (Ruby, 1984; 1995). Some are simply photographs within a small frame and covered with glass inserted into the memorial. Others are placed in frames and stand free on the monument or plot. The frequency of use varies between countries and regions; it is hardly ever seen on English gravestones, and is rare on Pennsylvania German stones (Graves, 1988:80) whilst it is more common on the small numbers of surviving late 19th and early 20th century monuments in Spanish and Italian cemeteries.

Angels became common elements on cemetery monuments in the late 19th and early 20th centuries (figure 21, right), and some were also allowed in churchyards. They were nearly all made from marble, and most were produced in Italy at centers such as Carrara, and exported around the world (for examples see Gillon, 1972; Brooks, 1989; Weston, 1989). In Catholic burial areas, figures of Christ, the Virgin Mary and other large figures were also supplied. Where cemeteries have expanded as they filled, such monuments are often clustered in certain zones of the cemetery, creating a dramatic skyline but often detracting from the impact of any one memorial.

Full scale figural sculpture of the deceased or the mourning family was not common because of its cost. It occurs rarely in Britain, but is slightly more commonly found in Continental European and North American cemeteries. Many items were again of Carrara marble, and were carved in Italy using photographs of the deceased. Such large-scale representational art was therefore not part of popular sculpture, but are found either as single figures, or on larger compositions. Sometimes figures are found outside mausolea, looking as though they are mourning at the tomb.

In America quite a number of figural sculptures are of children, of which some of the finest quality examples survive in Chicago (Brown, 1994:160–163) and Forest Hills, Boston (Wilson, 1998). Any form of personal representation was rare in 19th-century Australia, apparently because of the lack of sufficiently skilled craftsmen, though Robert Hancock had a larger than life figure carved out of local sandstone in 1844, and placed on his grave after his death in 1876 (Mackay, 1989:38). Small angels and cherubs, again often in marble and bought as stock catalogue items, are common throughout the English speaking world on children's graves in the later 19th and early 20th centuries. Lambs are also extremely common, often carved onto the tops of small headstones (Brown, 1994:155, 185).

Symbols of craft occupation occur from the 18th century (McCormick, 1979; Willsher and Hunter 1978) and throughout the 19th century, but are often only found in any numbers in a few regions. Membership of societies was also indicated by symbols, notably of the masons (Zaniello, 1980) but also other groups such as the Woodmen, and these are discussed in Chapter 7.

As remembrance became the major theme of texts (see below), so flowers came to dominate the repertoire of later 19th and early 20th-century memorials (Hobbs, 2002; Mytum, 2002e). These may be placed in a central panel, could spread across much of the top of a headstone, or be even more extensive and extend down the sides of the text panel. Some flowers were symbolic in themselves, but others were selected to demonstrate variety and richness of design. The numerous (though often not consistent) Victorian conventions of messaging through flowers may have been invoked by some memorials, but it is now difficult to know what carried specific meanings, and to whom (Hobbs, 2002).

8. TEXT

The lettering of text on memorials becomes much more standardized in the later 18th century, and shifts to an almost exclusively centered text. In the later 18th century large numbers of fonts could be used on a single stone, particularly on fine-grained materials like slate, and this continued into the 19th century in some areas. Gradually, however, standard font types came to dominate the carvers' repertoires, and lettering was no longer strongly expressive; the message carried by the text was emphasized only by lettering size and the use of italics and lower and upper case. Ethnic minorities used their own scripts, but Latin and Greek in the later 18th and 19th centuries was frequently used on monuments to the highly educated.

The introductory phrases that emphasized remembrance follow on from the mortality phrases, though there is a long overlap when both kinds are in common usage. The most frequent remembrance introduction was 'In memory of', but later this becomes elaborated and the emphasis is clearer. The phrases 'In loving memory' and 'In affectionate remembrance of' become the accepted introduction in the late 19th and early 20th centuries (Mytum, 1994). In strong contrast to the mortality phase, the use of the word 'died' became less popular by the late 19th and early 20th century, and euphemisms replaced it. These included 'departed this life' and at the end of the period 'fell asleep' became popular. The implications of these changes are discussed in Chapter 8.

Epitaphs tended to shift from warnings to the living from the deceased in the grave to explicit statements by the living. These either celebrated the virtues of the deceased, or emphasized the emotional state of the bereaved at their loss. At this time, many epitaphs were in verse. Whilst some were original compositions, a great many came from books from which an appropriate verse could be found to express the family's sentiments; most were of no literary merit, but clearly expressed the feelings of the bereaved. Quotations from the Bible were frequent, and could be several lines long (Mytum, 1999b). In many cases the Biblical book, chapter and verse is stated, but for others its origin was assumed to be known; in some cases only the reference is given and the reader was expected to know the verse. Lines from hymns and prayers were also chosen. It is noteworthy that whilst the sentiments implied by the decoration and symbolism of the stone often matched the epitaph, this was far from always the case. This may reflect a partial understanding of the symbolism or a mixture of emotions surrounding the death.

9. CONCLUSIONS

Over the period there is the rise to the greatest level of investment in burial customs and memorialization, reaching its peak in the later decades of the 19th century. By this time in many areas a majority or at least a significant minority were commemorated on a memorial. The density of monuments within burial grounds mushroomed during this period. What had been an open area with scattered mounds from recent burials and the occasional monument of the most affluent, became a built-up environment where row upon row of memorials were relieved only by areas occupied by pauper burials. The more affluent zones in some cemeteries, such as those of the rural movement in America, had areas where the plots were more spacious, but other parts of the same cemeteries reflected the cold commercial facts of supply and demand. Particularly in urban areas, cemeteries came to reflect in the management of death the realities of life, and the regimented and tightly packed homes in which the relatives lived and from which the deceased had come were replicated in the rows of graves and monuments.

Chapter 4

A Marginalized Activity

From After World War I

The dominant view amongst those researching attitudes to death, burial and commemoration is that the massive losses during World War I caused a dramatic and substantial shift in popular culture (Cannadine, 1981; Jalland, 1999). The social and ideological importance of the funeral, mourning and commemoration were all reduced, and this is reflected in the material evidence both above and below ground. There was undoubtedly a trend of this kind, though the amount of research to prove this causal link to material evidence has been limited beyond those studies of war cemeteries and memorials (see below). Alternative views also exist that need further consideration. One highlights existing attitudinal trends such as the growing interest in less complex ceremony and the rise of cremation in Britain in the later 19th century (Leaney, 1989; Tarlow, 1997; 1999b). Another view emphasizes the commercial pressures regarding the management of cemeteries, evidenced by development of the memorial park cemetery before the United States was involved with World War I (Sloane, 1991:155–163). Whatever the causes, and no doubt there were many, the funeral industry and burial and commemoration provision underwent as many changes in the 20th century as it had in the 19th. Despite a comforting aura of continuity exuded by the relevant professions, the combinations of commercial, social and ideological pressures through the century have left their mark.

One source of information for this period which is worth locating, if possible, are the trade journals in which new products and ideas were advertised and discussed. These include *The Monumental Journal* in Britain and *American Cemetery*, *Cemetery Beautiful*, *Modern Cemetery*, and *Park and Cemetery* in North America. There are also publications deriving from annual conferences of the various professional organizations linked to the funeral trade and cemetery management, regulations from cemeteries and church authorities, and advertisements in journals and newspapers. Oral history and contemporary attitudes derived from interview are also very important for this period, and have been applied with varying degrees in many studies noted here. Some useful anthropological research has been conducted on village and urban deathways and cemeteries in Spain (Douglass,

1969) and Italy (Goody and Poppi, 1994), and most notably in Greece (Bennett, 1994; Danforth, 1982; Kenna, 1976; 1991).

An important trend which affected social attitudes to death and commemoration was the gradual rise in life expectancy in industrialized societies, particularly noticeable with regard to the deaths of infants and young people (Lynch 1995). There has also been a rise in the proportion of the population that is elderly, giving greater impact to early death, and deferring the popular experience of the death of a loved one to a much greater age (Jupp and Walter, 1999; Sloane, 1991). Moreover, changes in media coverage and visual accessibility has led to greater awareness of death elsewhere, such as in war and famine, to the impact of accidents and terrorism, and to the deaths of famous people. Examples such as the assassination of President Kennedy, the Challenger accident, or the death of Princess Diana, created popular grieving and sense of loss has become more frequent if only for a passing time in each case (Kamerman, 1988:82–87; Walter, 1999). Begun in the 19th century through newspaper reports and memorial photographs of famous people, the sense of communal loss has become much more important in the 20th century. This has often been represented in behavior, such as the vast sea of flowers left outside Kensington Palace following Princess Diana's tragic death (Jupp and Walter, 1999: pl.21; Walter, 1999), as well as in monumental memorials.

1. THE FUNERAL

During the 20th century there have been very great changes in the timing, process and experience of death. Technology has created mass destruction in war, and dramatic reduction in contagious diseases, at least in the developed world. Changes in popular beliefs and the greater mobility of population has also affected all aspects of deathways.

1.1 The Context of Death

The trend for death to happen away from the domestic setting, noted in the later 19th century, was much strengthened in the 20th century. Hospital deaths became normal, and improvements in intensive care led to redefinitions of death, as bodily functions for some could be maintained indefinitely (Jupp and Walter, 1999). Control over life and death, and interest in quality of life and the wish for some to allow euthanasia, created a professional interest in death from groups as diverse as theologians, lawyers, human rights activists and doctors. Death itself, whilst supposedly a taboo, became a major topic of conversation and debate, whilst the actual direct experience of witnessing death declined rapidly.

Increasing institutional rather than family care for the elderly led to the creation of the nursing home environment where the old were placed where other age groups only visited, and from which death was the only escape. The hospice movement was another element in this trend. The concerns regarding death led to an increased number of autopsies, and donations to medical science meant that parts of bodies could be retained for research, teaching or organ transplants. This social distancing and further removing of death from normal life experience meant that the funeral industry developed methods which helped grieving relatives maintain that distance whilst still being able to make their own farewells (Habenstein and Lamars, 1981; Jupp and Walker, 1999; Sloane, 1991). Bodies were now normally held in a chapel of rest rather than returned to the home before the funeral; these chapels could be separate chapel-like structures but were often rooms within the funeral parlor with a vague, non-denominational quasi-religious style of décor, and allowed the viewing of the corpse by individuals or groups.

In North America embalming maintained and strengthened its position as a normal part of body preparation. Though found elsewhere in the world, it is here that it has been most pervasive, and evidence for it has been found archaeologically (e.g. Sledzik and Hunt, 1997). Other forms of body preservation, including mummification and freezing, have also developed in America (Quigley, 1998).

1.2 Changes in Organization

The funeral profession became subject to market forces and local firms were absorbed into larger organizations with buying power over coffins and fittings, and during the later 20th century providing pre-set packages. Whilst this had been the pattern offered by the 19th-century undertakers, this was marketed in ways that had not been attempted before. Whilst some objected to this modern version of commodification, it was popular with most people because it made making choices regarding the funeral like those of buying a new car or package holiday, and in that sense familiar and unthreatening. Pre-need purchases of burial plots and memorials were encouraged in North America (figure 29), and to a lesser degree elsewhere, and the old form of burial club was revived in insurance plans, often tailored to provide a certain specified category of funeral package (Sloane, 1991). Despite these organizational changes, however, many of the elements of the funeral remained at least superficially similar, encouraging confidence in tradition and respectability.

Elaborate funeral corteges and ceremonies were already in decline by the end of the 19th century, and were rapidly reduced in scale during the 20th century. The change from horse-drawn hearses to motorized vehicles hastened this change. Although specialist hearses were made, there was no provision for black plumes, large amounts of crepe, or easily visible undertakers and mourners. Ethnic differences in funerals continue to be strong, based partly on affluence but often linked to aspects of liturgy. The scale of attendance, the use of flowers, and the importance of ceremonies at places of worship, crematoria and the graveside vary considerably.

There is much discussion amongst the funeral industry professionals, anthropologists, sociologists and thanatologists regarding the effectiveness of the support given to those involved with death. These include those about to die, the bereaved, their carers, and including all those associated with the funeral experience, and this vast literature can be reached through journals such as *Mortality* and *Omega*. Some of this literature may be very relevant to particular historical archaeology studies, though the materiality which provides the context for emotion and action is often underplayed. It may be that archaeological expertise in modern material culture studies could be brought into supporting thanatologists in their endeavors.

1.3 Coffins and Caskets

In Europe, the single-break, flat-lidded six-sided coffin was the normal form and only a minority used a rectangular shape of casket. Moreover, most of these coffins are single-shell, and have a traditional appearance. In North America, however, the casket is the overwhelming choice of body container, and many have more than one shell (Habenstein and Lamars, 1981). This relates to a greater concern with preservation, already apparent with the preference for embalming. The sealing in of the body is much more important here than elsewhere. Most coffins and caskets continue to be made of wood, though veneered chipboard is normal in Britain and metal containers are found in America. Increasing awareness of green issues has encouraged the use of cardboard or wicker coffins (Natural Death Centre, 1996).

The dating of excavated coffins of this period is made much easier with the availability of many trade catalogues that provide the industry with the necessary fittings. Sales catalogs for the public also display the finished items, with the range of trimmings and prices.

1.4 Choices in Body Disposal

Earth burial was still the most frequent body disposal method in the 20th century, though Mediterranean interments in family vaults or communal *loculi* continued (see below). The use of family vaults in the rest of Europe has declined, though some established families have still continued the tradition. In North America, the construction of concrete vaults was common, and they were also made in iron (Sledzik and Hunt, 1997), yet a further example of the need to enclose and control the environment of interment. Alternatives and their attendant changes in attitude to the body are further explored in Chapter 8.

Cremation has given rise to a wide range of body disposal choices, and so have affected the nature of formal burial arrangements and the types of memorials, which are discussed as appropriate in the sections below. Other less formalized options exist for the disposal of cremated remains, termed cremains in America

(Sloane, 1991:228) but ashes elsewhere. They may be returned to the deceased's family in an urn to either retain in their homes, or dispose of in one of a variety of manners. A few common choices for informal disposal can be identified, though popularity varies by region, religion and age. Relatives can just keep the cremated remains at home or place them in their garden, or they may scatter them in a location they see as appropriate; popular places include at sea or some 'natural' place, or even at the deceased sport's fan's team stadium.

2. MOURNING

In some societies, such as traditional Mediterranean communities, the wearing of black by widows still continues, but in many contexts most of the elaborate mourning dress codes were rapidly abandoned during the early 20th century. Outward indications of mourning have largely disappeared, and individuals are expected to return to a normal pattern of life rapidly after a family death.

Decoration of graves is still a widespread phenomenon, and bridges the gap between mourning and commemoration; activities in the first years following bereavement are discussed here, longer term activities are discussed under commemoration. The most common is the placing of flowers on the grave as part of the immediate mourning process. In some regions natural cut flowers are used, or plants placed in a container or directly into the ground. In other regions, particularly those that are hot, artificial flowers are normally used, though even here appropriate plants and even shrubs may be selected. The tending of the grave, cutting the grass short or keeping it vegetation free, is also a frequent practice. The length of time that graves are visited varies according to cultural context and population mobility, but may not be for many years and even then only when a spouse survives. This has large implications regarding cemetery use and maintenance regimes.

3. COMMEMORATION

Non-material commemoration has become more common in the 20th century, with bequests to charities and institutions, and also in more practical material forms. The pattern of paternalistic, named giving has continued from the 19th century, now supported by governments (to varying degrees) through tax concessions. Whilst many gifts are anonymous, others carry the name of the deceased, and whilst the large gifts allow the construction of named buildings or endow substantial foundations, smaller amounts create memorials as diverse as prizes offered by academic societies to cups awarded at local shows. Many of these memorials have been self-selected by the deceased in their wills, and many are arranged before death. Only a minority of these types of commemoration are chosen by relatives after death.

Figure 22. Military cemetery area with small uniform tablets, many decorated with flags. Brockport, New York.

Activities within the burial context provide the main opportunity for commemorative practice, where the maintenance of graves provides the focus. Where there are large monuments, such as Mediterranean tombs, marble can be regularly washed, or plaster or cement painted. In other areas, tidying vegetation is the priority. In many regions, All Saints Day is an annual opportunity to carry out individual grave renovation and decoration. As many members of the community are in the burial ground on the same day, this is both a family-based activity on their plot, and a communal experience, visiting friends, sharing memories and, in some cases, carrying out work on the shared parts of the burial ground. In Louisiana and Alabama the annual grave maintenance provides an important opportunity for community reinforcement, and can include picnicking and religious services as well as physical effort (Florence, 1997:31–49). Other seasonal grave elaboration takes place; the placing of wreaths and Christmas decorations on graves happens widely in North America, even on the otherwise uncluttered memorial park cemeteries. This habit is now beginning to appear in Britain. Chinese families can gather at *Qingming* when they 'sweep the grave' (Chow and Teather, 1998, 289).

Another annual ritual is the placing of small flags on the graves of American war veterans on Memorial Day (figure 22). In other parts of the English-speaking world, such remembrance activity is focused around communal wreaths and individual small wooden crosses at war memorials. This activity started after the World War I, with the construction of war memorials. In Britain red paper poppies are worn in the days leading up to Remembrance Sunday, a symbol derived from the flower which colonized the battlefields of Flanders.

4. BURIAL GROUNDS AND CEMETERIES

Numerous locations are available for study from this period, and whilst they are usually easily accessible and relatively well maintained they are also often in active use, which affects how and when research can take place.

4.1 Location

The diversification in burial location noted in the last period has continued to widen, with a breakdown in the controls of formal religion and in the desire of individuals to have greater choice as consumers. By far the greatest change in location of burial has related to cremation, though until recently this was not an approved method of body disposal by the Roman Catholic church, and thus did not develop in some countries. Other options have also developed, however, including wider practice of private burial, attempts to preserve bodies in perpetuity, and new locations of burial such as woodland cemeteries (Constant and Waern, 1997). New cemeteries have been established on the edges of towns and cities, though in some areas such a change of land use is resisted by local residents, and planners can have difficulty in finding burial provision. The shifting pattern of names given to cemeteries in America reflects attitudes to such spaces and their associations with death (Zelinsky, 1976; 1990). The creation of war cemeteries where the deaths occurred has also led to the dispersal of commemoration across the globe. Thus, for example, the Commonwealth War Graves Commission manages burial grounds on every continent in cemeteries ranging in size from only a few interments to those of many thousands.

4.2 Intra-site Spatial Arrangement

Most family and church burial grounds have continued as before, though with the need at intervals to expand into new areas for additional burials. Infilling of plots where only unmarked graves had previously been sited delays the need for existing burial space, but most communal burial grounds in thriving communities have had to expand or, if there was no adjacent land available, a new burial area has been established elsewhere.

Many urban churchyards became full and were closed in the 19th and early 20th centuries, and for a while their importance remained when relatives of those interred there were still active. However, the need to widen roads as traffic levels rose, and the perceived undesirability of poorly maintained areas in the heart of towns and cities led to many such burial areas being developed. Some were built over by roads or other structures, and many became managed as small urban parks. This often led to the moving, lying flat or complete removal of memorials, replanting of vegetation and the placing of seating (Mytum et al., 1994). Many burial grounds now threatened by redevelopment and in need of archaeological excavation only became built over in the later 19th or 20th century. Today development is less likely to be the fate of burial grounds because of increased interest in the historic landscape and changing official attitudes to burial areas, Such cases that are proposed often create considerable media interest.

The regimented rows of graves developed in the 19th century continued in many burial grounds and cemeteries throughout the 20th century. With fewer burials in high density pauper plots, cemeteries expanded rapidly in size, and new areas were laid out to maximize numbers of plots. Segregation by religion was still strong in some cemeteries, but in an increasingly secular society mattered less for many clients. One of the major problems in well established cemeteries was the continued maintenance of old burial areas by then full. These were often left to become dominated by self-sown vegetation and, although for some this has created valuable woodland spaces in the urban landscape, it has often created havens for vandalism and crime (Dunk and Rugg, 1994). In contrast, areas of the cemeteries still used for burial and grave visiting are well maintained.

War cemeteries gave architects and designers an opportunity to create new landscapes of commemoration, with central architectural features surrounded by the ordered rows of identical individual memorials. Such cemeteries had developed in North America following the Civil War, but now they could unfortunately be created from the mass death and misery of World War I and later conflicts (Borg, 1991; Bourke, 1996; Ferguson and Turnbull, 1996; Mayo, 1988; Robin, 1995; Winter, 1995). These cemeteries have, because of institutional support, remained well maintained and conserved, and have often increased in their attractiveness as the architecture and planting has matured; they are now subject to considerable tourist visiting.

New civilian cemeteries and burial grounds have been laid out with present and future needs in mind. The problems that arose from the older management regimes have been noted and attempts made to avoid such difficulties. One of the management strategies that evolved during the early 20th century and has been popular in North America and Australia, but less so in Britain, was the memorial park cemetery. Developed first at Forest Lawn, California (figure 23), this style of cemetery played down the individual grave and its marker, and concentrated on a landscape enlivened by works of art and buildings of architectural distinction. Graves were marked by stone or metal plaques set flush with the ground, enabling the visitor a view of a rolling, contrived naturalistic landscape, whilst making grounds maintenance easy and inexpensive (Sloane, 1991). The adoption of this style of cemetery spread eastwards across North America from California, and

Figure 23. Left, Statue of Moses in the bullrushes and Right, Commemorative seat and typical cast commemorative plaque, both at Forest Lawn cemetery, Los Angeles, California.

a new generation of cemeteries was developed, once again on the edges of the suburbs where sufficient land was available.

In America, some community cemeteries were enhanced as work projects for the unemployed in The Great Depression (Ambler, 1998). Here, cemeteries without facilities such as chapels were provided with them, and improvements made to cemetery entrances and planting schemes. A mixture of styles was applied, depending on local tastes, but all choices were within the repertoire already present in cemeteries.

Crematoria have often been the financial salvation of cemeteries which have become largely full, and have led to a new range of body disposal and memorialization (Sloane, 1991). Besides the informal possibilities discussed above, burial authorities provide a range of options for disposal of cremated remains, though at any one location the choice may be limited. The ashes can be scattered at the crematorium, perhaps with the option of the addition of the name on a communal memorial. Many crematoria have a garden of remembrance where remains can be buried under a shrub such as a rose, and with a small commemorative plaque. Burial in a traditional family grave is also common, and can allow a plot closed to inhumations to continue in use. One formal method of disposal away from the crematorium is burial in a churchyard, now popular in Britain. This has led to the reuse of closed churchyards as cremated remains do not legally constitute a body (Mytum et al., 1994). Churchyards and cemeteries accepting the interment of cremated remains have a separate area set aside for them. Cremation memorials may be individual small plaques set in the ground, small headstones, or a communal memorial (Stapleton and Burman, 1988). The most elaborate commemoration associated with cremations are specially constructed cemetery buildings, *columbaria*, working on the same principle as Mediterranean *loculi*, though with smaller niches. These can be elaborate structures, and have given new visual foci to the cemetery landscape. They are most common in North America but are now appearing elsewhere.

In those European cemeteries where limited tenure allowed the reuse of ground, physical expansion has not been so necessary. In these cases, a gradual replacement of memorials of one generation by those of another has meant that successive zones of the cemeteries have become redeveloped and taken on a new character, with only the mature tree planting and path structure remaining. There have been two factors affecting the design and size of European cemeteries, however. The first factor was that of communist control of eastern European states, and the changed priorities in commemoration this engendered. Large monuments to the military were often erected, and areas were set aside for party members (figure 24)

Figure 24. Rows of identical concrete block memorials, Ljubljana, Slovenia.

even though those with religious convictions still retained their separate burial zones. There was also some redesign of the cemetery space to give greater access and visibility to the areas and features linked to the party and army. The fate of these historically important features now often remains in doubt. The second factor is linked to increasing affluence in western Europe, leading to a greater requirement for burial spaces due to increased purchase of permanently owned *loculi*, family vaults and earth burial plots. As these are not reused, the demands for more space, already long known in the English-speaking world, are now being felt in much of Continental Europe. At cemeteries such as those in Granada, Spain, parallel blocks of back-to-back *loculi* continue to be erected, a densely packed city of the dead. In other parts of the Mediterranean, rural cemeteries have been extended with more of the expensive family vaults on offer to cater to the affluent demand.

5. INTERNAL MEMORIALS

During the 20th century wall monuments became smaller and simpler, and were often relatively small plaques only with text. Internal commemoration continued to be diverse within religious buildings, however, and maintained the traditions established in the later 19th century. Stained glass windows were still commissioned, with often derivative designs in the first half of the 20th century, followed by a shift towards more contemporary designs in the latter half. Changing liturgical patterns in churches of all denominations has led to different furnishing requirements, and many of these items have been obtained as memorials and often have attached plaques or texts inscribed upon them. They are, however, far less obvious as memorials and reflect the desire to be remembered through a utilitarian item rather than just a memorial in its own right.

6. EXTERNAL MEMORIALS

With growing affluence, an increasing number of families were able to select a permanent memorial. Despite a lessening interest in funerary ritual, commemorative practices flourished through much of the 20th century, though modified in some regions by the growth of cremation. Towards the end of the 20th century, the western world saw a fall in support for traditional religious views and with it a rise in alternative methods of remembrance. In contrast, eastern Europe experienced a revival in formal religious expression, often linked to the redefinition of national identities following the collapse of communism. These factors had an influence on commemorative practice, though this was no longer such a significant arena in which social strategies were played out. The material effects of all these changes have as yet received little attention.

Figure 25. Left, Highly carved wooden cross, StDogmaels, Pembrokeshire, Wales. Right, Ceramic heart with impressed inscription set on rustic base with built-in flower holders, Etherley, Co. Durham, England.

6.1 Materials

Wooden memorials continue to be selected in many burial grounds to the present day, now often being of cross form. These may be used as a temporary marker until the grave fill has settled and the memorial stone has been selected, inscribed and is ready for erection. In many other cases, the wooden cross is the only intended marker, perhaps for financial reasons. Such crosses were not always welcomed, but as they were all that the poor could afford they were tolerated, even if against burial ground regulations. This was in the case, for example, at Botany Cemetery, Sydney (Zelinka, 1991:61). Many such crosses are now anonymous, but others were provided with a small inscribed plaque which may survive. Whilst wood has traditionally been a low status memorial material, it has recently become more acceptable because of changing taste, sometimes reflecting a more 'green' attitude to death and commemoration. Some of these commissions can be highly decorated, emphasizing that cost is not the only factor in such choice (figure 25, left). It is likely that, with greater family mobility, the commissioners of such monuments do not envisage grave visiting for a period longer than the life of a treated wooden memorial in a temperate environment.

In the rural *camposanto* graveyards of New Mexico, concrete became the popular material from the 1930s, replacing wood as the dominant material (Warren, 1987), and this was also the case with African American memorials (Little, 1989; Vlach, 1978; 1991). Concrete is also used in the construction of many of the vaults in America and Europe, though it may then be covered with marble or other cladding or rendering. Concrete was also chosen as a utilitarian and uniform material for many identical memorials in eastern Europe during communist rule. Ceramic monuments were also produced in some quantities in some areas, often in a wide range of forms (figure 25, right).

Cemetery monuments have been created in steel or aluminum, reflecting modern art movements, but these have been rare. Only slightly more frequent have been sculptures in stone or cast bronze which have attempted something beyond the traditions established in the 19th century. Funerary art has always been extremely conservative and derivative, and this has largely continued through the 20th century, with rare exceptions such as the D'Ascenzo memorial by Pitassi in Pittsburgh (Prestiano, 1983).

6.2 Forms

Average memorial size reduced significantly from the 1920s onwards, as the large headstone types became rare and a new range of forms became popular. Some taller memorials such as pedimented monuments with columns, and obelisks, continued for several decades, and crosses were very popular till around the middle of the century. Other types such as tombs and ledgers were rarely chosen, and low monuments only remained popular for a decade or two. These forms are not discussed further here as the trends continue those from the last period. New developments came in the form of lower headstone styles, many of which were associated with kerbs. Indeed, some memorials were made solely of kerbstones onto which the inscriptions were placed. Generally monuments were designed for one or two individuals, not for multi-generational use. Small flat memorials, for cremations or for interments in lawn cemeteries (figures 23, left, 26), also became widespread. Small subsidiary headstones, tablets and scrolls, already in use in the latter part of the previous period, continued to be erected as subsidiary monuments on American family cemetery plots but here, too, the lower separate stones and kerbing became very popular. In Continental Europe, changes came also from shifting patterns in commemoration as well as the forms used.

Large sculptured scenes continued to be occasionally commissioned in Europe, though family scenes became rarer and individual sculptures of the deceased, or of religious figures were most common. These figural sculptures and indeed family mausolea are also present in adapted form in America, where the expansive nature of many cemeteries allowed for the construction of such large monuments. The Davis Memorial near Hiawatha, Kansas, produced in the 1930s, is an example of this genre taken to its extreme, with a series of figural

Figure 26. View of a lawn cemetery with rows of plaques and occasional larger monuments, Forest Lawn cemetery, Los Angeles, California.

sculptures showing John and Sarah Davis through various stages of their lives (Brown, 1994:90–94). Sculpture in many forms, and including a wide range of materials, occurs widely in African American burial grounds, though the significance of many of the designs is not clear (Little, 1989:129–131; Florence, 1997:155; Vlach, 1978; 1991).

6.2.1 Headstones

The most popular burial marker continued to be the headstone, though rarely of local materials. During the first part of the 20th century, monuments with a touch of art deco style became popular, with simple shapes and minimal decoration continuing through the middle of the century. In the later 20th century some new forms became common, notably the heart, and in much of Continental Europe and Ireland a range of asymmetrical angular shapes became popular. Some Continental earth burials have headstones with elaborate features including glass-fronted niches, which could contain religious images, lamps, photographs or personal artifacts.

Some ceramic memorials were popular in the early part of the century (figure 25, right), and the plastic nature of the material allowed a wide range of forms to be constructed which were often not of headstone shape but may be mentioned here (Veit, 1995). Many African American headstones were made of concrete during this period, cast in a mold and with text and decoration incised or impressed whilst the material was still malleable (Little, 1989; Rotundo, 1997). Inlaying with other materials such as glass was also common. The range of forms is extremely wide, covering all the shapes encountered in stone, though many are unique commissions with designs not exactly repeated elsewhere.

War graves can be found in war cemeteries near to the location of death, or in the burial grounds in the place from which the person had originally come. Most war graves are marked with standard designs produced by that country, with standard format of text and motif (Meyer, 2001). Some authorities have allowed alternative memorials to be erected by families on some war cemeteries, though in others the uniformity of physical remembrance of death and sacrifice has been maintained at the cost of personalized grief (Tarlow, 1997; 1999b).

6.2.2 Flat Monuments

Lawn cemeteries allowed most interments to be marked only by flat plaques so that the overall landscape quality of the cemetery was not marred by a vista dominated by many small memorials (figure 26). Made of stone or cast metal, these plaques were often a standard size, though some variation in motif and text style was permitted (Sloane, 1991:183–184). In South Carolina, concrete slabs became popular additions to burial vaults in African American burial grounds, into which were set marble or bronze plaques containing the text (Little, 1989:109).

6.2.3 Additional Elements

Footstones became less popular in the 20th century, and if the grave were to be defined, then kerbs were used. Headstones were often designed to be integral with a set of kerbstones, and the limited impact of their low height and angled face was offset by the effect of the whole plot being enclosed in white marble or a granite. The interior area could be grassed or concreted over, but was often then covered with glass or rock chippings. This again gave great opportunity for a colorful impression, with chippings in white, gray, green or blue. Some Mediterranean graves were paved in tiles.

Some patterns of activity, such as the use of flowers, wreaths, and other *momento mori* still continue (Rahtz, 1981; Florence, 1997), with the use of synthetic materials allowing for greater range of items, and greater resilience against the elements. Additions to graves in the form of flower vases and flower containers were common (which may be purchased with already displaying text such as 'Dad'). Use of potted plants increased, as did the seasonal planting of bedding plants on graves as gardening became more widespread with rising affluence (Mytum, 2003d). Other additions included toys and ornaments on children's graves, and souvenirs from pilgrimage on Catholic plots. Many of these additional items were often discouraged by burial authorities because of the grounds management problems they created, but are still very common.

6.2.4 Kerbs

Kerbs as a monument type of their own were a relatively short-lived phenomenon in the middle of the century. The kerbs contained the inscribed text, sometimes around all four sides of the plot. Individual graves in some African American burial grounds were also marked by kerbs (Little, 1989:127–128). These were often constructed with coping and concrete blocks produced for construction and landscape work. Some were decorative in themselves, or arranged in a decorative manner.

6.2.5 Mausolea

Elaborate mausolea remained in demand during the early decades of the 20th century in North America (figure 27, left), but new examples became less frequent

Figure 27. Left, Egyptian revival family mausoleam, La Metarie cemetery, NewOrleans, Louisiana. Right, family tomb, largely subterranean and covered with marble. The tomb is elaborated with an angel and chain fence with bollards. Note the elaborate grave decoration with bunches of artificial flowers. North Front cemetery, Gibraltar.

after that time. In contrast, vault type structures in the southern states continued to be very popular, though not elaborately decorated, as an intermediate form between tombs and mausolea (Nakagawa, 1994). Many of the 19th-century mausolea continued to be used by established families in all areas. In Mediterranean Europe, mausolea became much more popular as investment in family burial monuments was still considered worthwhile. Numerous small mausolea have been constructed in the late 20th century for the burgeoning middle classes in the wake of an economic boom following entry into the European Community. Some mausolea have been constructed in contemporary architectural styles, but most have continued the Neo-Classical format. Smaller family tombs partly below ground also occur widely in the Mediterranean, and these could be made more elaborate by the addition of large religious sculpture, such as a cross or figure (figure 27, right), on top of the tomb (Mytum, 1993).

6.2.6 Loculi

The Mediterranean popularity for *loculi* has been maintained in the 20th century, with a greater number of families able to afford permanent burial locations. This has led to the growth in the numbers needed, and so additional blocks have been constructed at many sites, either above ground or as public catacombs, a below-ground version of *loculi*. Marble and cast metal plaques seal the *loculi*, though during the latter part of the century the marble slab was often inset slightly and a glass front placed flush with the surface of the communal structure. The intervening space is then used to hold photographs and personal or religious momentoes. The equivalent 'oven graves' of New Orleans have continued in use, though they do not seem to have retained their popularity to the same extent elsewhere in their traditional American south distribution.

6.2.7 Columbaria

Structures similar to *loculi* but for cremated remains are common in North America and are becoming more popular elsewhere. Rooms set aside in crematoria may also be furnished with small name plaques which may be purchased as a memorial. In many cases the *columbaria* and other memorial forms are purchased for a limited period, though perpetual services may be available though at a much enhanced fee. These forms of memorial provide some sense of physical presence despite the small size of the individual remains. There is also a sense of community both of death and remembrance created in such environments.

6.2.8 War Memorials

It was with commemorations of those lost in World War I that communal monuments became very widespread in most of the western world (Borg, 1991;

Inglis, 1992a; 1992b; 1998; Moriarty, 1995; Winter, 1995). World War II and later deaths have often added to existing monuments, though new memorials were also designed and erected. In America, the Vietnam War memorial shows the continuance of this tradition (Hubbard, 1984). The concept of the memorial to the unknown soldier is one which developed with the massive losses from the World War I and continues to this day (Inglis, 1993; Ingersoll and Nickell, 1987).

There are interesting elements of individual and group patronage, conflicting emotions, social engineering, and cultural affiliations which can be teased out of the material evidence of war memorials, and their often extensive documentation (Mayo, 1988; Winter, 1995). There is considerable use of past regional monument styles and decoration, through revival styles, but there was also some innovation, particularly with large memorials at war cemeteries. The forces of popular choice operating at local level, both conscious and unconscious, are worthy of much more research. The recent 50th anniversary of the ending of the World War II and the need for conservation work on many memorials built after World War I has led to an increased interest in this subject, though as yet mainly from architects and historians. This should lead to some important publications in the next few years, and archaeologists should contribute their particular perspectives on this expanding research area.

6.3 Decoration and Symbols

For much of the 20th century, decoration on gravestones was restricted in many cases to limited architectural features or a single central motif. In the middle part of the century many of the popular monument types such as the small headstone and kerbs, the cross or angel left limited room for other decorative features. Some of the late Victorian fashions for sentimental doves, flowers and lambs continued, but were widened to include a range of Classically-inspired motifs such as wreaths or simple geometric shapes which were popular on contemporary architecture. Most of the designs were produced at quarry workshops and sent to masons to add lettering. This was in a limited range of fonts, and was often painted incised lettering or inlaid with lead (Crossley, 1991). The flat plaques of lawn cemeteries were limited in size and contained little opportunity for decoration, though they could include a small religious symbol, a flower or some more personal decoration. In general, the mid 20th century was the low point in decoration of memorials, the resulting limited variability offering relatively little opportunity for subtle analysis.

In the last part of the 20th century, changes in technology have greatly enlivened the decorative possibilities for gravestones. Cutting and etching unique designs on polished granite began with hand crafted examples, but in recent decades has become much more common. Whilst mass production of stone shapes remains, decoration by the mason using computer technology has opened up the opportunity for many more fonts which can be cut by machine, and the use of different images to be etched onto the stone, one method being laser photoengraving. Whilst most

Figure 28. Headstone with etched designs showing tractor and cricket match. Buttercrambe, North Yorkshire, England.

of those selected are stock items, the catalogue can be much wider, and also personalized images based on photographs or linked to a particular interest can be catered for (figure 28). Other features may be chosen instead of flowers, though they remain very popular. The alternatives found in recent decades are items deemed typical of that person by the surviving relations, whether a teddy bear etched onto a child's gravestone, or a motor bike on that of a teenager's. Whilst many designs chosen may not be of great artistic merit, and burial authorities have rules to prevent what they deem inappropriate designs and wording, there is at least a great flowering of popular taste exemplified through these monuments (Gradwohl, 1997; Meyer, 1989b).

Photographs on memorials continue to be popular with some ethnic groups, but many of the more recent examples are photo-ceramics. In some cases the photographs have been transferred into an etched design. The ability to convert photographs into an image, such as with granite etching or engraving, and with photoblasting, is leading to a greater use of portraiture on headstones, and enabling their appearance to be more prominent (Meyer, 1993a:10–11).

6.4 Text

A return to the introductory phrase 'In memory of' during the 20th century was part of a trend to make all inscriptions shorter but should not be taken to indicate

Figure 29. Headstone commemorating a married couple, purchased and erected prior to either death. Birth dates have been inscribed, but not death dates. Note the prominence of the family name and the parallel panels for each individual. Brockport, New York.

a lessening role of remembrance. Indeed, many memorials only give the names and dates of the deceased, but symbols on the stone show affection. In some cases, such as lawn cemetery or cremation plaques, this was due to their small size, but a only limited amount of text was often chosen on larger headstones and crosses.

The use of text to identify linguistic and religious affiliation was perhaps even stronger in the 20th century than in the 19th, as there was less central control over memorial texts and as immigrant and minority groups sought to emphasize their identities; these are discussed further in Chapter 7. As the monumental industry and cemeteries promoted the pre-need purchase of plots and memorials, especially in North America, the wishes of those to be commemorated were clearly articulated within the designs. Monuments could then be erected with just the death dates to be added later (figure 29).

7. CONCLUSIONS

In a period of decline in relative expenditure on deathways, increasing affluence has seen a wider popular involvement with permanent commemoration. Reduced interest in burial and commemoration also contributed to a lack of concern for old cemeteries and their management. This was widespread during the middle part of the 20th century, leading to massive decay and destruction of monuments and whole burial grounds. However, renewed concerns for burial sites and monuments in the late 20th century has itself set in train different patterns of preservation with regard to historic remains. This has included the enactment of legislation, the establishment of friends' groups, and the implementation of education policies (see Chapter 10). Conservation of monuments and landscapes is now often desired if still only sometimes fully achieved. Formalized reburial of human remains at the same or carefully selected new settings for interment shows a new respect for the dead, even when contemporary attitudes to death are no longer taboo but ambivalent. With the new millennium, even later 20th century burial monuments are perceived as an archaeological resource, but public policy and attitudes to contemporary burial are in a state of flux. It is likely that further major changes will come about in the years to come.

Archaeologists, with their temporal perspective and awareness of the subtle meanings and power of material culture, should be part of the policy making process in the development of cemetery landscapes and regulations, and integrate more generally with professionals linked to the developing subject area of thanatology. Historical archaeologists have become very aware of the need to link their research with the local community, to recognize the rights of diverse interest groups, and to make the links between past, present and future. The archaeological study of the 20th century is likely to be mainly concerned with above-ground evidence, but the archaeology of funerary buildings, landscapes and monuments can be integrated with many other concerns in cognate disciplines primarily concerned with the present.

Chapter 5

Production and Consumption

One of the themes that can be explored using burial data of various kinds relates to the stages of commissioning, designing, producing, distributing and consuming material culture associated with death, burial and commemoration. A brief consideration of funerary material including some discussion of coffins and fittings and mourning material is followed by detailed consideration of monuments where most work has concentrated, and where the hand-made nature of the artifact until recent times means that the dynamics of production and consumption have particular value.

1. FUNERARY AND MOURNING PARAPHERNALIA

The undertaking industry has been studied in several countries, so that the major developments in the associated material culture have been given some attention, though not by archaeologists (Habenstein and Lamars, 1981; Kselman, 1993; Litten, 1991). Particular attention has also been given to the details of major state funerals, for which there is copious information including much illustrative material (Litten, 1991). Given the critical contemporary comments on the funeral industry during the later 19th century, there is evidence of some consumer resistance to the expenditure and complexity encouraged by the funeral industry. It would seem, however, that in the time of emotional stress families tended to conform, though may at other times begrudge the expenditure. Archaeology will be an important source of information regarding the degree to which sales pressure was resisted in practice, and across what parts of the population.

1.1 Coffins and Fittings

The development of the coffin fittings trade has been given some historical attention, and in England production was concentrated largely in London until the mid 18th century when it shifted to the west Midlands centres of Birmingham and Wolverhampton (Church and Smith, 1966; Fellows, nd). Initial technical innovation first by punching and then stamping was followed by a long period of limited development (Church and Smith, 1966:622), though there were some later stylistic changes. Some of the coffin fittings have been discussed within wider metalwork production, for example furniture handles being used as coffin grips (Gentle and Feild, 1975: pls. 348–366). Some trade catalogues have been retained in museums and archives, and a few have been reproduced (APT, 1980; Columbus Coffin Company, 1882; Fellows, nd; Meridan Britannia Company, 1882; Romaine, 1976). English products were exported colonial markets including India, Canada and Australia (Church and Smith, 1966:622).

How the coffin fittings were used, and by whom, has not yet been well explored. Some contemporary sources suggest regional preferences in Britain and Ireland (Church and Smith, 1966:623, quoting Aitken writing in 1866). Apparently, the Welsh preferred black and white pins and fittings, much of the west of England and Scotland enjoyed the use of some gilt fittings, and these were also most popular amongst the Catholic population of Ireland. In contrast, London undertakers apparently limited the use of such bright items, preferring black. However, such generalizations may be of limited value, as the London evidence from Spitalfields and other sites suggests a more varied and sometimes ostentatious display in the capital. Consumption evidence will therefore have to come from archaeology, though work is still largely at a preliminary stage. Published typological studies are still few, such as that for the almshouse at Uxbridge, Massachusetts (Bell, 1990), though some theses have considered the subject for America (see Bell, 1994). The producers'catalogs provide an indication of what was available and when, but not how the products were selected by funeral directors and then clients, and in what frequency. Only funeral directors' account books or archaeological excavation would reveal such important information. It may still be unclear how much detailed influence customers had in the selection of coffin fittings and their style, however (Litten, 1991).

The large sample of dated coffins at Spitalfields, London provided some useful preliminary results, but only a limited amount of analysis was undertaken (Reeve and Adams, 1993:83–89). Further research on the material by the author suggests that the rate of stylistic change in coffin furniture was much slower than that of commemorative monuments. The scale of this conservatism and the reasons for this are still being considered, but it is likely to reflect the different social role of the coffin in the funeral ritual compared with the subsequent monument. The different rates of stylistic change should also probably be linked to the different relationships between producer (or at least supplier) and consumer in the context of a funeral on one hand and a commemorative monument on the other.

The small assemblage from the Weir family burial ground at Manasses, Virginia can be used to show the potential for archaeological study of coffins, their fittings and the interments. Changes over time were identified on this site, with no ornamentation on coffins of the 1830s, and with the use of handles and decorative white screw caps only on the burials from the 1850s. Brass tacks (pins) appear from the late 1860s, and these coffins also have more decorative metalwork (Little et al., 1992:407–411). There was also a cast iron coffin, and two wooden coffins had glass viewing windows. The changes are explained by the trend towards the beautification of death, though there may be more specific factors relating to the family history which also played a part (Little et al., 1992:411–415). Only when larger dated and contextualised samples are available will the various factors of generalized trends and particular circumstances be more clearly distinguished.

1.2 Mourning Paraphernalia

Mourning clothing and other items survive in museums (Morley, 1971), but contemporary illustrations and photographs may give better information on the degree and extent of mourning within society. Etiquette books give much advice on the matter for the middle classes, but the practical materiality is better examined through financial accounts in family papers and probates (see Chapters 2 and 3). Those studying the production of textiles, costumes, ceramics and jewelry should be aware of the specialist products offered to the mourning market.

2. COMMEMORATION

The manufacture of many commemorative items such as printed cards or ceramics can be studied within the context of those industries and the ways in which unique or limited production items were created. Some other commemorative items were personal creations, such as samplers (Clabburn, 1998; Gabel, 2002), sketches or paintings, or photographs of the deceased (Ruby, 1984; 1995). Photographs could be taken whilst the deceased were alive, or as idealized post-mortem poses. By far the most numerous and durable example of commissioned commemorative material culture available for analysis, however, are the monuments in burial grounds, and this is the subject of particular attention here.

2.1 Materials and the Process of Monument Manufacture

Almost all early memorials in all areas are made from local materials, even if these were not ideal. This is one of the major factors in great variation in regional survival of memorials from this period. Normally stone or wood was used, though in some parts of Britain and Europe, some internal and a small number of external memorials were made of cast iron (O'Neill, 1965; Willatts, 1987). Amongst the highest status memorials, particularly interior monuments but also exterior tombs, a wide range of exotic materials could be utilized, but these merely emphasized the difference in status of those commemorated. Often the elements had already been carved on these imported pieces, as was the case with tombs and gravestones probably imported from England to Virginia (Crowell and Mackie, 1990; Mackie, 1988) and Charlestown (Combs, 1986). Examples of later 18th- and 19th-century memorials being exported to other colonial contexts have also been noted, for example in Gibraltar (Mytum, 1993).

During the late 18th century the changes in the scale and sophistication of the funeral industry has been noted. Memorials continued to be made in local materials, but superior sources such as slate and marble became popular. A New England example of this trend is demonstrated through the work of Thomas Gold of New Haven, Connecticut, whose early work is all in the local sandstone. In the 1790s, however, over half his identified output is in imported materials, slate and marble (Williams and Williams, 1988:54–55). Whilst the choice of imported stones varies slightly in different regions, there is a trend towards marble, including the global export of Carrara marble from Italy, followed by a rise in the use of various granites.

In the 19th century a wide range of materials became available, though they varied greatly in popularity. These included cast and wrought iron, terracotta, and other ceramics including imitation stone such as Coade stone (Kelly, 1990) and glazed products in a range of colors. These may have been used for complete memorials, as for the New Jersey monuments in a wide range of styles noted by Veit (1995) in New Jersey, or as elements of composite monuments. A notable example of the latter was the production by Jonathan Harner of finely detailed small terracotta panels to be inserted into headstones and some other external monuments in the Sussex area of England (Remnant, 1962; 1964). Few metal materials have received detailed study, though the examination of products in zinc (known in the trade as white bronze) of the Monumental Bronze Company highlights the possibilities of such investigation (Rotundo, 1989). Whilst the number of quarry sources increased with global marketing as the 20th century progressed, the use of other materials declined.

There is much discussion on the role of paint in gravestone appearance. Most paint, if used, has weathered away, and on the older stones none survives. In New England, Ludwig (1966:337) considered that early 19th-century stones may have been painted, and this deduction is supported by documentary sources and evidence from elsewhere. In the addendum to West's paper on the John Dwight workshop, some early 19th-century documentary evidence is presented for the preparation of an oil base for use on gravestones, but unfortunately no mention is made of the colorants added prior to application (West, 1989:27–29). Within Britain, there is much better evidence for the use of color, both in the 19th century and later, but perhaps representing earlier traditions. Coloring could be used just on capital letters, on all of the incised inscription, or also on the decorative elements. In some regions the whole face of the stone could be painted; some slate stones in Wales appear to have been painted white, perhaps to give the impression of marble

Figure 30. Mason carving additions to a memorial in situ, Ljubljana, Slovenia.

from a distance, or to overcome the somber dark gray. Stones may be repainted at various times, so that an existing color scheme may not have been the original one. In Ireland, stones may be sandblasted to remove lichen and make the stone look fresh, after which lettering and even designs may be painted. Many later 20th-century memorials use paint or gold leaf to enhance the visibility of text on polished granite, and the use of colored, etched designs has also become popular, often produced by a machine from a repertoire of designs held on computer.

Research on the working practices and workshops of funerary masons has taken place in a few instances, but could be much further developed (figure 30). Ethnographic study and interviews can throw light on production processes and the conscious role of ideology, Rotundo (1997) studied the production of concrete memorials by Merry E. Veal carried out in a semi-commercial setting, and Little (1998) records two African-American memorial makers. The accounts of working practices by masons themselves are also valuable; Crossley (1991) relates the process of letter cutting based on his career as a stone mason, and Brown (1989) considers his family business at Rookwood Cemetery, Sydney, over several generations. Printed catalogs and personally produced pattern books (Buckham, 1999) may give clues as to the range of products offered by a workshop, and annotations on such books can be particularly informative. The contents, with valuations, of the workshop of a Virginia mason Charles Miller Walsh, who died in 1901, has been published which shows the range of stock carried, and the tools necessary to carry out the trade at the turn of the century (Briggs, 1990:165–167). Other such documentary sources undoubtedly await scrutiny and publication.

Specialized production of particular forms can be noted. The distinctive rustic monuments made from Indiana limestone were first made about the middle of the 19th century but were widely popular from the 1870s to the 1930s, when they can be found all over America (Ridlen, 1996; Roberts, 1985; 1988; 1990). As was the case with many 19th-century monumental masons, the work of Indiana producers was largely standardized, though they also carried out other stonework (Lucas, 1990). Many designs and motifs can be paralleled in other forms of masonry, plasterwork or woodwork in a region. Only by the later 19th century in many areas was there sufficient work for stone masons to concentrate solely on memorials. Where specialists appeared earlier, they often worked on a particular material for which they had special skill, such as slate, marble or granite, and produced items such as fireplaces, sculpture and architectural moldings. Within this repertoire, memorials could take their place.

2.2 Production and the Role of Carvers

All early memorials, and many produced up to the present day, are worked by carvers and masons in direct contact with their customers. Although books of designs appeared from the late 18th century in England, these were then copied and adapted locally, rather than acting as sales catalogs. During the 19th century such books were widespread, and could reflect national trends. In the 20th century, many monuments were shaped and at least partially decorated before shipment to the local mason, leading to greater standardization in memorial forms over time. Nevertheless, local decisions were made regarding many memorials at a time when other popular material culture was almost exclusively mass-produced. Crossley (1991) recounts his working methods during the 20th century, but does not discuss his relationship with clients. Some memorials, such as those by African Americans, could incorporate mass produced items such as tools, mirrors and concrete building blocks, but placed together in unique and culturally significant ways.

The localized production stage of memorials has often remained within small workshops. The role of assistants and apprentices is little understood for the 18th-century memorials, but it is certain that the larger workshops such as some in Boston, Massachusetts, used such staff. Slaves may have been involved in production, but the evidence for this is limited and has been not universally agreed. The possible presence of an African American carver in Newport is discussed by Luti (2002:297–300). The organization of workshops, rather than the production of individual masons, has not been much researched, even in contemporary or recent settings.

In contrast to the local production of memorials, there were from the 19th century specialized firms established which were either where particularly suitable rock was available, or were at good distribution centers. Examples would include the Vermont Marble Company in Proctor, Vermont, Nichols & Company of Chicago, marble carving firms at Carrara, Italy, and granite works in Aberdeen, Scotland. Whole monuments, or elements such as railings, which were in metal were often made at specialist firms, and catalogs were published in many countries for these products. Nevertheless, most customers would only have known about these through local masons, who may have added these products to the range on offer, and taken a commission for ordering them and erecting them on the grave.

Little (1998:27) notes that pre-cut monuments were offered by large marble quarries and works, with instructions for local masons on the ways to finish the monuments to standard styles. These became even more widely available in North America with the 1902 Sears & Roebuck catalogue from which carved monuments could be ordered, with the lettering chosen by the customer added before dispatch (Little, 1998:27–28).

2.2.1 Identifying Specific Carvers

Some basic studies have been undertaken on monument seriation by archaeologists, but art historians have examined changes in form at a much more refined and contextualized level through a concern with the recognition of carvers and their products. The methodology of this type of work has been rarely explicitly stated, but it is essential for appreciating the patterns that can thus be identified, and from which many processes can be examined.

An early outline of the method is provided by Slater (1976), interestingly by a scholar professionally active in entomology. His approach includes elements of quantification not seen in most other studies, but in essence defines the generally applied methods. In order to identify a carver it is necessary to examine a memorial and note features not only of the memorial shape and size, but also treatment of the design elements in considerable detail, though the same design elements may be used by more than one carver (Brown, 1992). The form of lettering as well as what was said may need to be recorded. If all memorials produced by a carver had the same features it would be easy to identify the products, but of course there is variation at any one time and indeed over time due to a range of factors. These variations increase the potential interpretive value of the studies, but there is not always a great deal of discussion given in publications as to why the changes are interpreted in a particular way. Thus, even in Slater's great study of Eastern Connecticut memorials (1996a) there is, despite his scientific training, no introductory methodological section.

The omission in the literature of a detailed exposition of methodology has been partly addressed in the detailed analysis of the 18th-century carvers in Newport, Rhode Island (Luti, 2002). Here the methodology is described and illustrated in some detail and, for those wishing to undertake carver analysis this is a valuable guide. Lettering styles can be amongst the most characteristic features of a carver, and Williams (2000) has indicated which letters, because of their frequency in monumental inscriptions and their potential for variation, are those which should be considered first.

The art historical approach is based on direct field observation and recording of traits and through the use of high quality photographs and/or rubbings. Though it is not described in these terms, it is clear that the method of analyzing the memorials is an example of a hermeneutic dialogue between researcher and the memorials, with constant reference back and forth between models of classificatory ordering and the stones themselves. The factors that become critical may be statistical analysis of motifs, but more often and more effectively it is usually the subtle treatment of particular features. These features may be subsequently explicitly described and illustrated, but they come out of the data after a prolonged period of appreciation and absorption by the scholar.

A fine example of the how various elements can be defined and seen in numerous combinations are those of the Mullicken and Lamson family carvers in Massachusetts (Tucker, 1992; 1993). In the case of the first, the various motifs used on the side panels and finials defined the workshop, and the idiosyncrasies of the lettering related to the family member involved; with the latter, selective elements are defined for some of the products. Sometimes clear evolutionary trends can be identified, such as the ways in which John Huntington carved noses and eyes over a twenty year period in the third quarter of the 18th century (Shepardson, 1996:154).

Carvers can be identified when they sign their work. Unfortunately, this has always been a rare practice, particularly on exterior memorials. Individual carvers only signed a small proportion of their work, and even then not always in the same way. For example, of the 2124 tree-stump style memorials studied by Ridlen (1996) only 3% had the name or even initials of their mason. With families carving stones over several generations, and in areas such as Wales and Ireland where many people in an area had the same names, this can cause confusion. By the 19th century, it is not always clear what the signature implies. Tyson notes that it could indicate complete production, that it was made or lettered in the workshop of the mason, or even that the mason had been responsible only for importing and erecting it (Tyson, 1994:616,626). In cemeteries, it was often the practice to carve the company workshop name, in a range of abbreviations, and it may not be possible to identify individual masons in these circumstances, though each may have had their own preferred form.

Despite the various caveats outlined, there is no doubt that signed stones aid the art-historical identification of a carver's work, and provide an important link to other historical sources such as census returns, church records, and trade directories. Where a mason did carve a significant amount of work, what was signed may be seen to vary over time, and could include place of production if the monument was to be erected away from the normal zone of supply, as with Charles Lloyd Neale's largely unremarkable mid 19th-century products (Finnell, 1993:100). Some studies, particularly of 19th-century masons, but also sometimes on particularly well documented 18th-century carvers, have been presented in a biographical style, attempting to reach the person rather than just their products (Smith, 1994; Knoblock, 1996). Occasionally masons have recorded aspects of their craft (Brown, 1989; Crossley, 1991); others have been interviewed, though publication setting their work in context has been rare (Little, 1998; Rotundo, 1997).

Throughout the world, many cemetery companies have extensive business records that define grave spaces, indicate ownership, and list interments; some cemetery records provide some information regarding payments for memorials. Increasing numbers of cemeteries have been the subject of detailed histories. Though these studies do not necessarily cover all the aspects of interest to archaeologists, they may mention some masons and architects involved in monumental design and production (Linden-Ward, 1989a; Meller, 1985; Murray, 1991; Zelinka, 1991).

In eastern North America probate accounts often provide information on the commissioning of 18th-century memorials. Whilst many only note that a payment was made to a known carver, but without mention of its purpose which has therefore to be presumed, there are numerous examples which explicitly state that the carver is being paid for a gravestone. The records form a powerful tool for analysis in New England, and indeed the activities of many carvers can be explored using the probates, and can thus include many commissions which have not survived or which cannot now be identified. For example, in a study of seven early 18th-century Boston carvers, only 43 of their products are identified through initials on the monuments themselves; the remaining 56 are identified from the probates (Chase and Gabel, 1988). Probates have thus been extremely important in stone carver identification and dating, as without them most craftsmen would have no name and their work would only be tentatively assigned to a single worker. In Britain there is little documentary information available regarding most memorials, though some early memorials in Scotland can be linked to permission to erect them (Harrison, 1990).

2.3 Commissioning and Production of Monuments

When considering the process of commissioning and carving, and the relationships between producer and consumer, a range of sources can be used. These have been most fully explored for high status memorials in Britain and Ireland (Llewellyn, 2000), and for headstones in 18th-century New England via the probate record discussed above.

The high quality British memorials, largely internal but including some external products, are generally viewed in terms of the style and abilities of the artist

(Whinney, 1988), though there is sometimes recognition of the role of the patron. The headstones of New England have been largely seen as products of particular craftsmen, in a folk tradition (Benes, 1977; Ludwig, 1966; Slater, 1996b). There has, however, for some time been some concern with the role of the patron, though it has not been well developed (discussion by various participants in Harding, 1976:21–22). Instead, the carvers are seen as being the creative force from which the consumers chose an already produced design from a limited range available.

Two factors in consumer choice would be important and relevant in this form of analysis: the cultural norms expected in terms of stone form and design, and the costs of the memorial. Whilst some progress in understanding these factors can be gained just from the material culture, the most confident interpretations are those supported by other sources.

An important feature of probate records is that the charge to the carver is often given; thus an economic dimension can be considered using these records which survive in large numbers in New England. Where the documented stones also survive, costs can be compared with the actual product. In some cases, prices can be found carved on New England headstones, where they would be visible in the workshop but would be buried beneath the ground once erected. Where these are also probated, the prices can be compared. Those in the records are generally much higher, presumably because they cover all costs of carving, transport and erection. It is unclear as to whether the prices carved on the headstones are prices before the chosen inscription, but it is likely that this was the case. It would seem, therefore, that in some workshops a selection of prepared and priced memorials were available for perusal by customers.

The account and day book of Ithamar Spauldin covering a period over 1795 to 1800 when he was based at Concord, Mass, has survived (Jones, 1980). These records greatly increase our understanding of a particular craftsman and the dating of his products, but also have considerable implications for the operation of the craft in New England. A study of these records by Wilson (1992:108) demonstrated that similar sized stones were produced for men and women, but that children had smaller memorials, and that size also correlated closely with cost. That prices seem to be related to the size of stone rather than shape or decoration has been demonstrated elsewhere (Stier, 1983), though in the study of Ithamar Spauldin, it would seem that the wealth or status of the customer could have been a factor (Wilson, 1992:108). This suggests that more complex factors in negotiating price may have been important. Wilson (1992:107) also considers that a range of products were available to view, as headstones and footstones from the Spauldin workshop do not match in 49% of the surviving 53 known pairs, suggesting that choice was made from what was in stock at the time, rather than being carved specially.

Other examples of documents useful in understanding the commissioning and production process include account books and even diaries. The account books of Calvin Barber survive (Petke, 1993) which could allow considerable analysis, not only of back dating, (see pp. 180–181) but of who purchased the stones, and even when several were purchased at once. The diary of one carver has been studied to understand the context and method of production (Tucker, 1995).

The shift from cherubs to the neo-classical urns and willows can be observed in the general trend found in so many New England burial grounds (Deetz and Dethlefsen, 1965; 1967; 1971; Deetz, 1977), but can also be studied through the work of a particular carver. This allows for assessment and interpretation of the personal adjustment to common taste or the evolution of a carver's style (which was cause and which the effect has not been given the attention it deserves). One such carver, Calvin Barber of Simsbury, Connecticut whose work largely covers the decades either side of 1800, has been studied in detail (Petke, 1993). Carvers may also have a variety of styles available at any one time, and so appeal to a range of tastes, as with Jotham Warren of eastern Connecticut in the late 18th and early 19th century (Slater, 1996b).

In Britain and Ireland probate records are earlier than the period of graveyard memorials under discussion here, and so there is less direct evidence for the commissioning of stones. Grogan (1988:59–60) considers the client base of Dennis Cullen in later 18th century County Wicklow, suggesting largely Catholic but also some Protestant customers, based on the surnames of the deceased. Some memorials are more useful for an analysis of the commissioning process in that they state the name of the person who had the stone erected. This is frequently the case with internal monuments, but tends to be a regional fashion on external memorials. One study on memorials from County Louth in Ireland illustrates the potential information that these explicit statements of commissioning may provide (Mytum, 2002b).

In the 19th-century cemetery context, increased documentation of the burial process and the wider contextual evidence available from other sources, has encouraged a greater consideration of the role of producer and consumer. Clark (1987) assumes that the commissioners of memorials had considerable choice over the style chosen. Whilst this freedom could in part be constrained by functional factors such as cost, it was more conditioned by actual and desired degree of acculturation of immigrant groups into what was seen as mainstream American culture. The details of the decision making process remain obscure, however, unless other sources of information are available. Pattern books are frequently noted for the 19th century (Burgess, 1963), and whilst some are printed and were widely distributed, others were hand drawn and can be used to consider in more depth the negotiation of memorial choice by the mason and client (Buckham, 1999).

The larger mausolea have attracted much attention since many were constructed by well known architects and artists, and their funerary work can be set within a wider context (Brady, 1993). But this approach does not usually consider the client (Wright, 1988), though there are exceptions. In the case of Thomas

Crawford's design for Amos Binney, Dimmick (1992:166) assumes that the surviving wife played a role, and for larger commissions such as the mausoleum at Castle Howard, there is no doubt that the client was a significant force in the decision making process (Saumarez Smith, 1988).

3. TEMPORAL CHANGE

The classic and much-quoted studies by Deetz and Dethlefsen (1965; 1967; 1971) provided models of gravestone change over time in New England. The battleship curves produced show very clear rise and fall in popularity of various motifs, the three main design items being the death's head, cherub, and urn and willow. Other evidence from New England reinforces this pattern, and the work of individual carvers can show the shift from one form of design element to another over the lifetime of the craftsman (Blachowicz, 1998; Petke, 1993; West, 1989; Williams and Williams, 1988). This demonstrates that change was not brought about through generational replacement but by alteration in output through a working life. Welch (1987) provides data for the change over time of the death's head and the cherub soul effigy for Trinity churchyard, New York, and usefully also includes plain stones which not only Deetz and Dethlefsen but many later studies have underplayed.

Studies of change over time away from New England have shown the rise and fall in popularity of forms and motifs, though rarely with the same clarity as in New England where particular social expectations may have limited choice. In the 19th and 20th centuries, a wider range of forms can be noted (Cannon, 1989:441; Mytum, 2002e). Moreover, the choice of individual motifs, the ways in which they could be combined, and the materials from which the monument could be constructed were all much greater. A broader social and ethnic background of the client base also means that the battleship style of presentation does not necessarily provide the best way of displaying the complexity of the data. However, it has been used to good effect where an appropriate level of simplification allows general trends to be identified (Dethlefsen, 1981).

Many archaeological studies of temporal change assume a pattern of diffusion of forms and motifs over time that gradually leads to an increase in popularity of an attribute, only for it to decline as another attribute replaces it in popularity. Indeed, from the early studies in New England, this trend has been noted (Detz and Dethlefsen, 1965; 1967), and then further identified elsewhere (Gorman and DiBlasi, 1976; Dethlethsen, 1981). Several studies emphasize an emulation process by which elite designs are copied in smaller or less elaborate forms by the aspiring middle classes and, by the end of the 19th century, sections of the working class (Cannon, 1989; Parker-Pearson, 1982). Cannon (1989) offers some quantified evidence for this diffusion being one of emulation in Cambridgeshire, England, with features being taken up by the higher status groups and only reaching the monuments of agricultural laborers significantly later. By this time, other features had become popular for the elite. In contrast, Tarlow (1999) argues strongly against emulation in the development of monumental commemoration, and considers that this is generally a weak form of explanation. It is worth noting that emulation implies simple copying, whereas instead the deliberate taking on of features of form and style may be to create new and different forms. These may also incorporate features not derived from higher status models, and even those so derived may be used in a very different way and with completely different social and symbolic purposes. This debate is one worthy of much further attention, and needs studies based on careful contextual analysis. Little et. al. (1992:415) also note that it is desirable to augment processual form with what they term cultural content in order to understand the significance of emulation in mortuary behavior.

With elite internal monuments, the patterns of innovation and imitation can sometimes be better understood because of the surviving documentation between craftsmen and clients that indicates sources of inspiration. This is often assumed for external monuments where, for the early monuments, the book by Quarles (1635) is seen as a direct inspiration for design motifs though ideas would also have come indirectly through other illustrations in a range of media that had in turn been inspired by such books (Burgess, 1963:167–168; Ludwig, 1966:274–183; Willsher and Hunter, 1978).

4. SPATIAL CHANGE

The distribution of carved stone memorials because of limited local supplies has been noted above, but studies of the areas covered by particular mason, workshops or monument forms are also valid approaches. A few studies have also considered distribution of burial ground types.

4.1 Regional Studies

Deetz and Dethlefsen (1965) modeled a Doppler effect in the spread of popularity of style across the Massachusetts landscape during part of the 18th century, suggesting the way in which gradual cultural change could be measured. Whilst interesting as a document of the New Archaeology, it did not provide any insight into the reasons for this, and assumed it was somehow more widely applicable. In the period since this pioneering work there has been an enormous amount of research into the products of carvers, the influences on their styles, and the distribution of their works (Blachowicz, 1998 and other references above). Many studies have tended to produce lists of burial grounds where the products can be found, often with the numbers of examples and sometimes details of the dates. For example, the detailed study of the Connecticut carver Thomas Gold has produced 645 examples of his work from 24 burial grounds. However, the extent and nature of this distribution (less than one third of the output was erected in his base, New Haven) is merely noted, and not discussed at any length (Williams and Williams, 1988). Although changes in popularity of the work of various carvers has not been numerically modelled, it is likely that much more complex sets of factors would now be identified which affected the spread and take-up of products of particular masons, and the spread of design elements between them.

In some contexts, such as along the eastern seaboard of America, movement of suitable stone for memorials began between colonial centres at a relatively early date. Either raw material or completed monuments could be transported. The carver could be near the source of stone and export memorials, or be near customers and import materials; Solomon Brewer began his career at the source, and later moved to the demand (Williams, 1994). Most early imports were of completed memorials. Surviving examples include stones from New England and Philadelphia in Charlestown (Combs, 1986), whilst St. Mary's County, Maryland was supplied from Philadelphia, then Baltimore and Washington (Mackie, 1988). Long Island received stones from New England, New York, and New Jersey (Levine, 1978), and Nova Scotia obtained headstones from New England up to about 1780 (Trask, 1978a). North Carolina obtained some high status memorials from England, and others from Massachusetts, New York and Connecticut (Little, 1998: 29–68).

In Nova Scotia it has been possible to show dramatic shifts in memorial source over time. Early memorials were imported from Massachusetts Bay, but after 1780 were made locally. Scottish influence through immigrant stonemasons was also a factor in Halifax, though one identified carver, Abraham Seaman, came from New York (Trask, 1978a; Trask and McNabb, 1988). Other shifts caused by the movement of craftsmen have been noted within New England.

A few comparative distributions of different monument types within a region have been produced. The study of 105 southern Ontario burial grounds combined the results of all these sites for most aspects of the study, rather than investigating variations between sites (Norris, 1988). However, the distribution of the semi-circular topped headstone over space and time was considered in terms of urban, urban margins and pioneer fringe. The isolated areas were only slightly slower than the urban in their adoption of the form, and Norris (1988:145–146) notes that even here the time when this form was most popular predated trends identified elsewhere. Regional studies of monument typologies have been produced in other areas, such as Pennsylvania (Hannon, 1989), Wisconsin and Oregon (Francaviglia, 1971), and Louisiana (Nakagawa, 1994). In Britain the relative popularity of different monument forms have been assessed in north Pembrokeshire, Wales, across a range of denominational burial grounds (Mytum, 2002e).

4.2 Distribution of Carvers' Products

The works of particular craftsmen in Ireland have been plotted by Longfield in a series of short papers (1943; 1944; 1945; 1946; 1947; 1948; 1954; 1955) and the County Wexford carver Dennis Cullen has been reassessed by Grogan (1998). Grogan also updates Longfield's overlapping distributions of known carvers in south-eastern Ireland with a new map of known stones (Grogan, 1998:63). Some initial work has also been undertaken in the north of Ireland (Smith, 1993). Distributions of particular early monument types or decorative features within a region have been attempted in Britain, in Cardiganshire (Chater, 1976; 1977), the Cotswolds (Elliott, 1978) and west Yorkshire (Brears, 1981). They have been plotted to show areas where particular regional styles existed; all these examples do not represent one carver but a wider participation in a set of stylistic conventions. Such localized patterns continue through the 19th century, even in the face of imported stones and nationally developing styles, one example being the pedimented stones of west Wales (Mytum, 1999a).

Distribution studies of later products have been rarely attempted. The Indiana limestone monuments are an exception in that the range and extent of the products have been evaluated (Roberts, 1985; 1988; 1990), but distribution maps and quantified studies over space or time have not been produced. The problem with some of these later period products is that they were produced on such as scale, and distributed so widely, such quantification would be a very large undertaking. Some studies in America have looked at the output of local masons and give some idea of the sphere of activity of such firms. Examples include that of Charles Miller Walsh of Petersburg, Virginia (Briggs, 1990), Nathaniel Holmes of Barnstable, Massachusetts (Blachowicz, 1998), and Charles Lloyd Neale of Alexandria, Virginia (Finnell, 1993).

4.3 Intra-site Patterns

Spatial change within a churchyard should be a relatively straightforward aspect of analysis, but even this has been rarely attempted. Most studies have concentrated on the typology of memorials rather than the spatial relationship of one with another. The development of the topography of a churchyard was first explored at Llangar Church, Wales, by Shoesmith (1980), and a study which covers the evolution of a graveyard up to recent decades has been undertaken at Kellington, England (Mytum, 1996), and for an 18th-century graveyard at Killeevan, Ireland (Mytum and Evans, 2002).

Distributions of excavated burials have received some attention. The density and locations of interments within churches in England has been noted at a number of sites including Sevenoaks, Kent (Boyle and Keevil, 1998), Wharram Percy (Harding, 1987), and St Mary-le-Port, Bristol (Watts and Rahtz, 1985). The density of vaults is notable at some urban churches such as St Augustine the Less, Bristol (Boore, 1998), and much attention has been paid to the location and movement of coffins within the communal crypt at Spitalfields, London (Reeve and Adams, 1993). Work on English churchyard burial of the last few centuries has been hindered by limited excavation and intensive reuse, though some Quaker sites of limited duration show well order burial areas (Stock, 1998b). At the Kingston upon Thames site, however, there were numerous north-south and east-west burials which may reflect different phases of site use (Bashford and Pollard, 1998).

North American burial grounds have tended to be used for shorter periods of time, so the layout has often been clearer when excavated. Most burial grounds of all periods show care in the ordering of graves, though this was less structured in the case of early farmstead burial areas (Gibb, 1996), and with much more attention to rows of graves in larger burial grounds, though again relatively few are usually closely dated.

5. CONCLUSIONS

Whilst the general trends associated with funerals, mourning and commemoration can be set out (as in Chapters 2–4), the process by which decisions were made, and the forces which directed change over space and time are less well understood. The association with named and dated individuals for many burial monuments and, in the most comprehensively studied cases, also the relevant interments, allows the possibility of carrying out far more detailed analysis than is often the case in archaeology. The commissioning and production of mortuary material culture is one where the historical and archaeological sources are both complementary and necessary to increase our understanding of the dynamics discussed here.

Chapter 6

Social Structures

Mortuary analysis in archaeology has often concentrated on the social implications of the disposal and commemoration of the dead. Historic cemeteries provide rich evidence for social structure, from excavation but also in a more accessible form through memorials (Moore et al., 1991). The content and style of memorial texts, the arrangement of monuments within burial grounds, and the choices made over size, form and material, can often be linked to social structures. Tensions between social groups and between the owners and users of burial grounds can also be identified and analyzed. Many social changes through time and space have a correlate in the memorial evidence, and more importantly it was sometimes in the arenas of burial grounds that such tensions were actively played out.

1. THE RELATIONSHIP BETWEEN LIVING COMMUNITIES AND BURIAL GROUND POPULATIONS

Social structures are not monolithic and easily defined, and often communities of the living do not directly correlate with the assemblages of memorials or burials within burial grounds. Some studies, however, have been optimistically simplistic regarding the relationship between the population commemorated on the burial markers and the living population of a community (Foster and Hummel, 1995). Many factors affect the population reflected in the archaeological evidence, and these need to be considered before further analysis and interpretation can take place.

A major influence on the representation of a living and dying population is that a large section of the population was unable to afford a memorial of any kind. Many markers were uninscribed, and could be of any local material such as field stones or timber (Jordan, 1982; Mytum and Evans, 2002). These tend to be difficult to date and could easily be reused on a number of occasions, so they are frequently less well considered in any analysis. It should also be remembered that some that could afford a permanent inscribed stone only chose a memorial of a perishable material such as wood, or did not feel that any grave marker was seemly. Relatives visiting graves would know where they were located, whether there was a marker or not, so the need for a permanent memorial was not necessarily as great as might be assumed by modern researchers. Some denominations also did not accept the need for grave markers; for example, the Quakers did not officially permit grave markers until during the 19th century (Stock, 1998a). In more recent times, those who are cremated and have their ashes scattered may leave no memorial, or one of a form quite unlike that marking an interment. These might take the form of an entry in a book of remembrance, or be of a temporary nature, such as a wall plaque that requires a additional payment after a short period of time.

The frequency of memorial use can be assessed through a comparison with burial registers, if they exist, and if loss rates of memorials can be estimated. Together, these sources can indicate the degree of representation at a burial ground, and can give information about the extent of infant and child mortality (which is usually under-represented in memorials before the 20th century, when by contrast it may become over-represented), and causes of deaths.

Some groups desired markers which cannot be read and interpreted as easily as an inscribed stone, such as African American populations in Texas who use lengths of metal pipe and cactus plants to mark graves. Others chose a different form of memorial such as an item of furniture or a stained glass window within a church or chapel, or some form of charitable foundation that bears their name. Thus, researchers need to be aware of alternatives that might cause the population represented by burial ground memorials to be only partially representing any past living community.

2. STATUS

Much effort has been expended on considering the role of status in mortuary archaeology. Measures have often been taken on the basis of materials used and range of grave contents and memorials structures. These are often quantified in terms of energy expended, though the style (involving expertise and therefore quality of what is produced within any unit of time) does not seem to be so often considered. Where prices are known, from probate inventories or catalogs for example, the financial implications of different choices can be studied with some precision. Where some items are of known prices, others may be compared and have a cost inferred. Though changes over time will have affected costs, it would seem that in many cases location was more of a factor in supply than actual price before the late 18th century when larger scale production and distribution began to affect the memorial profession. Thus, more isolated communities may have just had less choice and so found the demonstration of status more difficult, rather than not being able to afford differentiation.

Historic burial ground data falls into two distinct classes from the point of view of status analysis. Where much is known about the status of the individual, the study can concentrate on comparing the material form and expenditure on commemoration and that status; in the situations where status is unknown, then the study may be concerned with inferring the status from the material evidence. In some regions, with partial survival of sources, analysis of the first kind may be possible on enough individuals to create a framework from which the rest of the surviving memorials could be evaluated. The material evidence itself, in the form of the memorial inscription or the name plate on the coffin, gives the vital information regarding name, age, sex, and other contextual information. This in many cases can be further augmented and expanded using documentary sources. For example, Richmond (1999) indicates burial register entries in Glasgow, Scotland, which demonstrate the scale of the funeral and its cost for a range of burials in the 1830s and 1840s.

In some cases, status may have to be inferred on more generalized principles. This applies to many unmarked graves and excavated remains without independent information on the individual. Even where some personal information is available, from an inscription, for example, this may not give any indication of status and other sources may not be easily available. Whilst some inferences regarding status may be appropriate in these circumstances, it is clear from documented cases that the scale and elaboration of memorials may have little to do with status. Whilst resources have to be available to pay for the memorial, unless it is home-made, this may be paid for by a benefactor such as an employer or wealthy relative, or as a communal action, such as was frequently the case with memorials to religious leaders. Moreover, some of the more elaborate mausolea found in cemeteries may reflect wealth over status, and could be used to help create the aura of an established social position which may not have been accurate (figure 31, left). Some burial locations, such as in Westminster Abbey, London, automatically confer a certain degree of status. In such a location this was further elaborated by the choice of form, decoration and detailed placement of the memorial.

Vaults of higher status families have been the subject of examination and in a small number of cases, exhaustive analysis (Butler and Morris, 1994; Caldwell, 1976; Litten, 1999). Some vaults have shelves along the sides to house the coffins, in a manner similar to the Continental European mausolea, but others have coffins resting directly on the floor or slightly raised on timbers (Mytum,1988); sometimes coffins were stacked up on top of one another (Litten, 1999). There is little evidence for differential status accorded to the location of coffins based on their occupants, though the actual coffins themselves, their fittings, and funeral acoutrements such as wreaths and crowns, can indicate social status (Litten, 1991; 1999).

Figure 31. Left, Elaborate mausoleum in a prestigious location to a showman, Kensall Green cemetery, London. Right, Family tomb with primary central commemorative panel and many smaller plaques subsequently added, La Fayette cemetery, NewOrleans, Louisiana.

3. FAMILY STRUCTURES

Many aspects of familial relations can be investigated through funerary monuments. Below ground study can be of family vaults or excavated graves, where the arrangement of interments can by highly instructive. Excavated human remains can also be studied to identify traits which suggest familial relations, though in this section concentration is placed largely on memorials, as the opportunity to study significant assemblages of burials is limited.

3.1 Scale of Family

The family unit varies greatly, both in terms of the number of those living together, and the composition of the group which feels a sense of belonging. The family structure can be multi-generational and it can include siblings and cousins. The family in death can be seen to vary over time and space, from extreme individualization to extended multi-generational families. This has been described briefly as non-dynastic and dynastic burial (Fenza, 1989).

Family burial can be examined in terms of the space used by one familial unit for interment. This may be a plot within a communal burial ground, discussed below, or can be a discrete isolated burial ground solely for the use of one family. Such sites can be found adjacent to many isolated colonial settlements, though they are rare in Britain and Europe. They show little differentiation in treatment of burials and often are remarkably limited in variability in memorials, save general changes over time, though the founding patriarch may be given a more substantial monument, around which smaller subsequent markers may be placed.

The size of memorial often gives an indication of the expected number of people to be commemorated; modern memorials are often small, whether for inhumations or cremations, and are sufficient for one person, or for two people if these are a couple. In contrast, later18th- and 19th-century stones can be much larger. During the 19th century, the symmetrical layout of text restricted the number who could be comfortably commemorated on one headstone often to around eight people. Some designs split the inscribed face into two long, narrow panels onto which more individuals could be inscribed in text of a relatively small size. In the later 18th century, where running texts and split words allowed the denser concentration of text on the headstone face, up to perhaps a dozen people could be commemorated on one monument, especially as epitaphs and long accounts of individual achievements were rare. With larger and more complex memorials, various surfaces could be inscribed; chest tombs were the most generous, with all four sides and the top providing five surfaces for inscriptions. In the late 19th and early 20th centuries, the densest pattern of commemorations can be found on crosses with stepped bases, where each step offered four sides for inscriptions. Early 18th-century and older headstones are notable for having few individuals commemorated by name on most stones, a feature seen on the earliest stones in Britain and Ireland, and New England.

In Europe, the desire for continued use of the same memorial led to the addition of small inscribed plaques to the sides of chest tombs to increase the number of people commemorated, often over a period of a century (Mytum, 1993). A similar process can be seen at New Orleans where some family mausolea have additional commemorative plaques attached to the walls (figure 31, right). This same long-term family association can be seen on Irish headstones where there is often a significant chronological break between late 18th- and 19th-century commemorations, and then mid 20th-century additions of quite a different style thereafter. Some Irish grave plots incorporate the historic stone with a new headstone to continue the commemoration (Mytum, 2003b; 2003d). The old stone may be placed erect at the head of the plot next to the new stone, or may be laid flat over the surface of part of the plot, visible and revered, but not detracting from the new stone by being noticeable from a distance.

In many European cemeteries, family plots are marked only with monuments containing lists of names with dates of birth and death, the relationship between the

Figure 32. Left, Headstone with kerbed plot infilled with stone chippings and decorated with artificial flowers. Note the additional commemorative plaques at the foot of the grave. Holyhead, Anglesey, Wales. Right, Children's graves with toys and small figures placed on and around the memorials, Amsterdam, Netherlands.

various individuals being unstated. Here the unity in death did not need justification or explanation; to the family it was understood and to others it was unnecessary. Such memorials usually have no individual epitaphs or descriptors of any kind, and it would seem that after death all that mattered was familial association. The family name is displayed in large letters on many of these monuments, a strategy also employed on Irish, Scottish and North American stones (figure 29). Many mausolea only display the family name, with the personal details of no individuals visible externally (figure 50, left).

In the 20th century, kerbed graves offered opportunities for multiple inscriptions, which could be further increased by separate features such as open books, vases and slabs with more inscriptions placed within the kerbs (figure 32, left). In cemeteries, larger plots could be set aside from the mid 19th century, with room for a range of memorial types set in rows, often around a large, impressive 'founder' burial marker. These can be seen in most American cemeteries where inscribed kerbs are rare, but the placing of numerous small, individual memorials together within a family plot is widespread.

The use of fencing and railing was an important characteristic of 19th-century cemeteries in particular (Linden-Ward, 1990), though it also occurred to a lesser extent in other burial grounds, when it began in the 18th century. The American fenced burial plots could be entered through an often elaborate gate inscribed or decorated with symbols (Forbes, 1990; Gayle, 1990), though in Britain and Ireland the gate into railed plots tended to be made in appearance like other sections of the fencing. Unfortunately most railings were removed from British graves during World War II, but a few survive and others can be studied from old photographs. Most were of a standard type used for fencing, and could be higher than many of the surviving American railings and a real deterrent to entry within the burial plot. Some plots in many cemeteries around the world were marked not by a vertical fence, but by bollards or posts in stone or metal, joined by chains. These provided a clear definition of the plot, but easy access. In many cases the chains have been removed but the original design can be inferred from the fittings on the bollards, or by fixing points for the uprights on the kerbing.

The normal pattern of inscription on memorials where there is not a long-term use of a monument is that of the married couple, possibly followed by some of their children who die in their parent's lifetime. There may also be unmarried siblings who are subsequently included, and occasionally more distant relations such as uncles or aunts, grandparents, or cousins. These are quite rare, however, and the nuclear family tends to be the main unit of commemoration.

3.2 Male Roles

Male social dominance can often be demonstrated in 18th- and 19th-century memorials, with a distinct shift to equality in representation during the latter part of the 20th century. The most obvious way in which this is represented is through the order of commemoration, where the male head of household is recorded first, even if he died after his wife. This demonstrates the point at which a permanent marker is placed over the grave, and this may be decades after the first interment recorded lower down on the stone.

Another method of emphasizing male dominance is to describe the wife in terms of the husband. Where the wife dies first, the reader may sometimes be excused for thinking that memorial is to the husband as so much is stated about his position. In many regions the occupation of the man is an important role to be recorded, as may other public offices held, such as mayor or justice of the peace. It is relatively rare until recent times for father, dad, papa or other expression to be used as descriptor after the details of a deceased male in the way that mother was used on female commemorations. Men are sometimes defined as fathers, however, when a child is described as son or daughter, and both parents are stated.

3.3 Female Roles

In contrast to men, the overwhelming emphasis on memorials to females is in relation to offspring. Whilst wifely virtues may be lauded, it is motherhood that is the dominant theme. When there is some epitaph it frequently draws attention to any motherly qualities and the grief of the bereaved children. Children who died at a young age are often only inscribed on the monument after their mother's death and as part of that episode of updating the memorial, further emphasizing their association with the mother.

Women are often described in terms of their personal character, particularly on elite 18th-century monuments and to a lesser degree on a wide range of 19th-century memorials. Christian virtues including charity, compassion, concern for the disadvantaged, good works, and widespread respect, are all worthy of note. Similar qualities are often still mentioned, though in less effusive terms, on 20th-century stones.

Descent is important in higher status families in particular, and this may be even more important to record with women who change their surname on marriage. Many women have their own family name mentioned in some way. This can be through the use of alias, as on many Irish memorials, or née, used in Britain. It may alternatively be by a whole descriptor phrase giving details of the woman's mother and father, even to the extent of giving the father's occupation, parental pedigrees, and place of abode.

3.4 Children

Children are normally treated in a subsidiary role in commemoration. Whilst there were always some families who were both affluent enough and thought it appropriate to erect a separate memorial for a young child, in most cases these early deaths are added to the memorial of their parents. Infants are rarely named and given an age, and may not even be separately acknowledged but are merely counted or even just mentioned in the plural. In the first few years of life, a child's age was often recorded by years and months, and this is reflected on the memorials. After the age of 5 years, the smaller divisions of time no longer serve a useful purpose. Those few infants recorded in detail may have their life measured in weeks, days or even hours. Studies of the changes in definition of young individuals in terms of name, age and relationships could be particularly informative, especially if this did not just concentrate on the few with elaborate memorials.

It is possible to see the development of a social persona through childhood that is reflected in the treatment on the memorial. Very young children have made little distinctive mark, and so any epitaph or decoration is generalized. In older children, the identification of characteristics and interests can affect the way in which they are remembered. Modern child's graves may have toys or scenes etched onto the headstone, and may be surrounded by toys which rapidly take on a forlorn and weathered appearance, a decaying symbol of time stood still for the deceased yet moving on for the grieving parents and family (figure 32, right). On earlier memorials there may be mention of educational achievement or particular virtues, the equivalent of the adult statements of occupations and character.

Children sometimes have full size memorials, but may have smaller ones. These are usually identical in form and material to the adult graves, but are to scale. Thus, small kerbed plots, some with railings, mark the size of the interment, not the full plot that an adult would have occupied. Headstones and crosses can also be small, but chest tombs and other types of more elaborate monument are almost always adult size even if for a young child. Not all small headstones of a particular form are for children, however, and further research is necessary as to why other individuals (not necessarily poor or socially insignificant) had small memorials. Children's coffins were necessarily smaller than those for adults, so they were provided with fewer fittings, and some of these could also be supplied in small sizes (McKillop, 1995). In England, children's coffins were often covered with white cloth, or painted that color, and so were clearly differentiated in appearance (Litten, 1991).

Children had particular motifs, notably lamb, dove or angel, on late 19th- and 20th-century gravestones in many urban cemeteries around the world, including Upland South folk cemeteries (Jeane, 1987:71). A study of 905 19th-century memorials to children in New Castle County, Delaware, gives numerical data on the ways in which families found consolation through the choice of monumental iconography and epitaph (Smith, 1987). Snyder (1989) has suggested that the innocence of children was sentimentally expressed on Victorian memorials, and Brown (1994:152–89) illustrates many evocative though not necessarily typical memorials to children in American cemeteries. Many children, however, were not commemorated by families on memorials, or frequently they were only remembered in a brief manner at the end of the text commemorating their parents. It may be that emphasis on memorials specifically for children skews the sample towards a tiny group highly emotionally affected and who wished to show this through monuments. Others may have showed their grief in other ways. The deaths of children in recent times has become rarer, and so reactions can be more extreme (Lynch, 1995). A far higher proportion of child burials now have memorials, and these are often the most decorated grave sites in contemporary cemeteries.

3.5 Family Relationships

When members of the same family are buried in adjacent plots or within one large family plot (figure 33), they can emphasize that relationship by selecting similar or identical memorials, or they can chose memorials that are very different and so demonstrate their individuality (Mytum, 1999a). Both these strategies can

Figure 33. Family plots, some with large founder monuments, Mount Auburn cemetery, Boston, Massachusetts.

occur in the same burial ground at the same time, and from the memorials themselves there is usually nothing to help explain why one route is taken by some and a different one by others. Documentary sources may help, but probably only family oral history could throw light on the motives and sensibilities behind such decisions. These sets of memorials do help to highlight issues of degrees of similarity, and what features should be the same to indicate familial proximity. The dynamics of stylistic choice can be exposed as very volatile yet subtle. The spatial arrangement of memorials belonging to a family can also be a productive line of inquiry (Brown, 1993).

Monuments may have been selected with reference to the deceased in terms of size, form, and decoration, as well as textual content. Two late 18th-century New Hampshire masons, Paul Colborn and John Ball, produced stones with decoration which seems to have been based on the age or sex of the deceased (Chase and Gabel, 1997:230–232), but it is rarely as clear-cut as this. One of the reasons is that stones were often designed to commemorate more than one individual.

The redefinition of the family can be examined through the declining mention of relatives and in the use of single generation stones, but can also highlight substitutes, notably pets (Kay et al., 1984; Meyer and Gradwohl, 1995; Spiegelman and Kastenbaum, 1990). The types of information provided for animals, and the degree of elaboration of the memorials, reflects their place in individual owners' lives, though not the wider community. Pet memorials were also erected by the British elite in their estates, though these were usually relatively small and plain memorials placed together in a secluded location.

4. INSTITUTIONAL STRUCTURES

Social structures within some institutions, and between their members and the rest of the population, can be explored through mortuary evidence. Pauper burial grounds often leave little trace, though an important study has been carried out at Uxbridge, Massachusetts (Elia and Wesolowski, 1989; Bell, 1990; Elia, 1992). Memorials may be simple or few, as in that case, or could be more numerous. Thus, military cemeteries have many memorials, though even here there is a tendency for conformity. Where this is not enforced, as is the case for example with some British military burial plots in commercial cemeteries such as Brompton Road, London, then the wider social stratification can be revealed beneath the attempts to show military death and sacrifice as equal amongst all ranks. One interesting group of institutional cemeteries are those created by particular companies for their employees, usually in situations where parochial or community burial grounds were not available, for example on tea estates in Bangladesh (Radford and Farrington, 2001). Some of the cemeteries show remarkable uniformity of monuments, whilst others show much greater variety, though the reasons for this have yet to be explored.

5. PATTERNS OF DEPENDENCY

Memorials were expensive items that for the 18th- and the first part of the 19th-century could not be afforded by working class families. Servants of the aristocracy and some wealthier clerics could, however, be provided with memorials. These family retainers could be honored with a memorial, though not without it being made clear both by the positioning of the burial and by the text on the stone the nature of the relationship between deceased and the provider of the monument. The same could be so for key estate workers such as bailiffs, gamekeepers and estate managers. In death as in life, the established order could be honored and perpetuated.

Most slaves were not provided with memorials by their masters, but that was the case with a small minority. At death, some may still have been slaves, and others freed, but it was their owners or previous owners who could mobilize resources for the headstone. Some studies have been carried out summarizing slave headstones (Krüger-Kahoula, 1989; 1993; Malloy and Malloy, 1994), and others have concentrated on the particular site of Newport, Rhode Island (Garman, 1994; Tashjian and Tashjian, 1989).

6. SOCIAL HIERARCHIES

The position of a burial, and of any subsequent memorials, was often heavily influenced by the social standing of the family of the deceased. This may relate to which burial ground was used, or where within the burial ground interment was located. Thus, social and racial segregation could be relevant, or cost could be a factor. As burial spaces became occupied, and as desirable locations for interment and commemoration changed over time, so the dynamics of social differentiation led to shifts in the competitive social arena of death.

6.1 Elite Burial

It had long been the case that a small aristocratic and religious elite was able to be buried separate from the rest of the population, whether in the most favored chancel area nearest to the altar, or in side chapels to particular saints. These remained important areas of burial for Roman Catholics and some high church Anglicans for theological reasons, but the rise in internal burial was largely been based on the social prestige derived from visible commemoration within the church.

Just as the elite separated themselves in life, so it was often the case in death. Within the context of worship, the change to seats and pews led to a stronger structuring of space, with the development of box pews, pews at different heights and degrees of decoration, and the addition of galleries to church designs. This physical representation and negotiation of social structure and social space became well developed within the religious sphere in Britain, Europe and North America. With the middle class burials spreading along the aisles of the nave, the aristocratic patrons constructed family vaults in the chancel of the church for which they were patron, or in the churchyard. This could be immediately adjacent to the church, or at a little distance. The role of the mausoleum on the family estate became significant from the 18th century onwards in Britain and Europe. In North America, the family burial ground on a plantation would be an example of the spatial separation of social groups, some of them being relatively wealthy. In urban centers such as New York, the British and European experience was repeated.

With the rise of the middling classes, the numbers of people who could afford and would aspire to a socially significant form of burial grew. This was reflected in the rapid increase in density of burial within churches, leading to much intercutting of graves and disturbance of human remains (figure 40), and the development of strategies such as brick-lined shafts and family vaults to limit this. The placing of floor slabs with inscriptions over or near the graves led to English and many Continental European churches being largely paved in such memorials. Wall space also became cluttered with memorials of various sizes, though these could be cleared during church restoration or rebuilding. Many such monuments, on walls and floors, have subsequently been moved, so that spatial analysis of the remaining memorials can be problematic. Where *in situ* survival can be demonstrated or reconstruction postulated, such as through antiquarian descriptions, the spatial and social dynamics of elite and middle class internal commemoration could be usefully explored. Few internal memorials would seem to have been erected in North America, but those that were would have performed the same functions as those in England.

6.2 Pauper Burial

Whilst most external communal burial grounds linked to churches might have allowed the elite some differentiation, all the rest tended to receive similar treatment. This may be visible in an historic graveyard by the intermittent use of memorials, with many graves unmarked, or by the presence of many field stones or timber memorials between more substantial monuments. This mixed pattern shows that burial of all classes took place in close proximity. Commercial cemeteries wished to maximize income, and so from the 19th century onwards many people were faced with choices regarding location and type of burial which was based on cost and, in some cases, other social criteria also. In this way a more complex spatial element in the social stratification in death could be constructed, and was a clear outcome of the rules and regulations of each cemetery. Urban churchyards under pressure from increased demand fueled by population expansion and immigration from the countryside usually had trenches dug which were gradually filled with large numbers of bodies (Curl, 1972; Etlin, 1984; Sloane, 1991). The restricted space in urban graveyards gave little opportunity for commemorative display with monuments by those who could afford it. Cemeteries in Britain, Europe and North America during the 19th century provided the middle classes with an arena for commemoration and remembrance, and the poorer sections of society with interment less unpleasant than in the churchyards. Cemeteries thus often had areas set aside for common burials, at a low charge and with no rights for a memorial (Barnard, 1990; Murray, 1991; Sloane, 1991). These interments were in graves that contained more individuals than private graves, and they were filled not with relations but whoever died over the short period of time that the grave was open. The material evidence for these burials may be very limited, but documentary sources should allow their place in the overall demography and mortuary culture to be considered.

7. NECROGEOGRAPHY

The social geography of burial and commemoration can be carried out through analysis of references to place of residence on memorial inscriptions, or through documentary sources which can locate the deceased when alive and compare with place of interment or commemoration. For example, the areas served by cemeteries in rural Ontario have been examined by Norris (1988). Data from gravestones was merged with tax roll and directory data to identify location of residence, and produced what have been termed "deathsheds" for Euphrasia Township, Grey County. Whilst most individuals were buried within two miles of their homes, some were buried at a greater distance, and not at the nearest burial ground. This was partly related to the denomination of the deceased and the nearest appropriate burial ground, and partly to population movement of the younger generation to the town of Meaford, resulting in parents being buried there rather than near where they had lived.

8. EMULATION

The normal assumption regarding the spread of cultural traits regarding mortuary behavior has been one of emulation from the top down, with the gradual assimilation of innovation first taken up by the elite then appearing amongst the middle classes, and subsequently the rest of the population, according to resources. There are undoubtedly elements of mortuary behavior, whether it be the introduction of coffins or the rise of intramural burial, which would support this model. The same can also be noted in the spread of permanent commemoration and in some of the monument forms and styles. There are many features, however, which do not comfortably fit this model. Tarlow (1999) has suggested that the tradition of external commemoration in Britain was developed apart from elite fashions, though this is far from clear in many regions. More obviously linked to popular rather than elite culture are ethnic traditions of burial and commemoration found in colonial contexts. The most widespread example of this would be African American commemorative traditions, which owe nothing from any elite culture.

Patterns of emulation, directly between carvers or indirectly through the use of pattern books, can be better appreciated as the products of individual carvers has become identified. This can now be frequently noted in areas such as 18th-century New England where the characteristic features of many carvers' work have been studied. For later memorials, recording can note features that would allow detection of the use of particular memorial pattern books (Buckham, 1999). Emulation is also recognized in terms of motifs copied from Emblem books and funerary stationery or forms from furniture and architectural detailing in the 18th century, and later from these and many printed sources. The use of many pattern books was not restricted to one medium, but they could be applied to products in wood, stone, or textiles. It could be used in a range of internal and external contexts; the links between design elements on needlework samplers and New England gravestones has been highlighted by Gabel (2002). She has noted how the death imp appears only on samplers and gravestones, whilst other motifs appear on a wide range of material culture. What is important to consider, however, is not only the source of inspiration, but rather the values, meanings and significance of the attribute in both its original and new context. There is little doubt that the 19th-century mason in particular could be extremely eclectic in taste, and clearly not to the dissatisfaction of the consumer. Thus, emulation could be social, ethnic or religious, and could be derived from contemporary or more ancient sources. Memorial art and design has rarely been innovative, but the manner, source and purpose of the reworking can be highly informative. This is where an archaeological and folk life perspective on the evidence is likely to be more productive than an art historical one where primacy tends to be given to the innovatory phase. In cultural terms, later dissemination and use of styles, symbols and motifs can be far more significant that the original creation of such features.

The nature and degree of emulation within and between social and cultural groups needs far more research than it has yet received, and so the overall role of emulation in mortuary practice and in memorial traditions is far from certain. As more detailed local and regional studies are completed, the patterns of connections will become clearer, though it will require considerable work to link memorials to other categories of material culture.

9. CONCLUSIONS

Social structure has been examined in many ways within the mortuary context, and it has been a subject of investigation by many scholars. There is, however, still very great potential for further investigation at a number of levels. The impact of large scale social change can be seen reflected in the memorials, but at a more local level the burial ground can be seen as an active arena within which social relations were played out.

Chapter 7

Identities

One of the most interesting and valuable aspects of memorials is that they are designed to encapsulate certain culturally important features of the deceased. The total text available, and degree of pictorial representation allowed, was restricted by practical constraints such as size, and by social convention. Priorities therefore have to be made in terms of what descriptors an individual will be given. This has already been assessed within the context of social relationships, but other aspects of identity may also be indicated, or may even supplant familial ties. These other kinds of identity can take two forms. Some identities are very conscious and are deliberately selected, to set the individual apart as a member of some group. Others are normative and subconscious within that group but to the anthropologist today, or indeed any external viewer, can be seen as identifying the deceased with a group. In many cases there is an element of conscious selection, made within a subconscious framework. There has as yet been little research on this balance. For example, Ridlen (1996) has briefly indicated how a particular monument form, the tree-stump tombstone, can represent any of six aspects of identity, through its form, additional symbols and text. She lists these as family and home, religion, occupation, patriotism, association, and agrarian values; most would seem to be conscious, but not necessarily expressed in those terms by those who commissioned the memorials.

All individuals have many personae which cannot all be elaborated on the memorial. The main categories are discussed below, but one valuable form of analysis, possible where there is good documentation, is to consider what aspects of individuals are not emphasized. Thus, for some the religious affiliation will be clear, for others not (though they may be just as devout); likewise some individuals may emphasize membership of a society, an honor bestowed, or an occupation, whilst others in the same burial ground may not. This can be seen with the graves of famous individuals who may have a perfectly normal memorial, such as authors Jerome K Jerome and Sir Arthur Conan Doyle (Greenwood, 1990). Thus, studies may in the future concentrate not just on what is said or symbolized (which is the case with the many examples discussed below) but what is not; we then more clearly understand the processes which lead to selection of certain identities in some circumstances and not in others. National identity is rarely emphasized, as it is redundant, though Clark (1989) has indicated evidence for an emerging American identity in memorials in Carolina belonging to Scotch Irish settlers in the later 18th century. Nationalism linked to ethnicity or war losses is more frequently indicated (see below).

Affiliation can be represented by plots, sections of a cemetery, or even by whole burial grounds. This representation can be in the form of choice of memorial but can also be through boundary representation, such as the fences around 19th-century cemetery plots (Forbes, 1990; Linden-Ward, 1990) which can be merely generic, or be specific through the use of symbols or text to define the grouping. The layout of a whole burial ground can also indicate affiliation, through orientation, features such as chapels or founder burials, and the nature of the burial ground boundary and entrance gates. Affiliation may indicated by the choice of location within a particular burial ground, be it to family, place, religion, ethnicity, or any combination of these. It is rare to have occupation as the primary factor in spatial arrangement, though this is the case at St Andrew's Ukrainian cemetery, near New Brunswick, New Jersey (Graves, 1993:55).

Some of the more common categories of identification on gravestones are listed below, though often the same memorials can indicate a number of identities, and the examples selected are merely illustrative. Some funerals, details of the coffin and its contents, and the interment itself, may indicate aspects of identity, but these are less amenable to study.

1. RELIGIOUS AFFILIATION

Religion is never more important within a family or community than at the time of death. The religious feelings of the deceased affect the treatment of the corpse, the funeral rituals, nature of body disposal and subsequent memorialization, if any. The location of burial in many cases adjacent to places of worship (whether churches and chapels used for everyday worship or cemetery chapels constructed specially for funeral services) is another factor which creates an atmosphere where religious identity would be manifest. It is not surprising, therefore, to find that many studies have focused on this form of identity, or have worked on other issues within a context of defining groups for study on the basis of religious affiliation.

Most of the burial grounds which are available for study have been created in a predominantly Christian milieu, though not all memorials in such burial grounds mark the burials of those with such a conviction. The most common early exceptions are small family burial grounds which may not indicate any particular religious affiliation, and Jewish burial grounds in some cities. From the 19th century onwards, burial became a matter more of public policy or private initiative through the development of cemeteries. This secularization of the burial landscape meant that those who wished to emphasize a religious identity had to profess this through the memorial, rather than just the place of burial, though many cemeteries catered for religious persuasion by zoning space by denomination. The very fact that cemeteries could be religiously mixed meant that such a definition could be seen as more important. Set against this trend, there was a move by the majority to no longer emphasize religious affiliation or belief at all in memorials. The changed environment of commemoration may have been such that this aspect of a persona, even if present, no longer seemed appropriate.

In this chapter, religious leaders could have been considered under religion or occupation. Their memorials often encapsulate many of the most prominent features of an association with religion, but given their special status they do not often form a useful basis by which to identify the mass of believers of that denomination. For that reason they are discussed under the occupation heading.

Published detailed comparisons between different denominations within a particular burial ground or a region have been surprisingly rare, though they have produced some valuable evidence. A comparison between Anglican and nonconformist burial traditions in north Pembrokeshire identified language use as a key variable, with material variation in popularity of certain monument types such as obelisks and crosses (Mytum, 1994; 2002e). In an Ontario sample, Presbyterian, Mennonite and Catholic denominations were present and showed distinct variations in popularity of styles and motifs (Norris,1988:142–143). Cunningham (1989) has specifically examined the differences between Navajo and Mormon graves in the Ramah cemetery of New Mexico.

1.1 Christian Denominations

1.1.1 Roman Catholic

Through both iconography and key phrases on memorials, it is often possible to identify Roman Catholic memorials (figure 34). Given the belief that prayer for the souls of the dead can reduce their time spent in purgatory, there are often phrases asking for such help on the memorials. Many begin with the phrase 'Pray for the soul of', or end with the Latin '*requiescat in pace*', 'rest in peace', though the shortened version R.I.P. can be used by other denominations. There may also be invocations to Mary, Mother of God, specific saints or saints in general, asking for prayers for the deceased. One also has to be aware that some high church Anglicans may use these phrases, though in almost all cases prayer for the soul is a Roman Catholic feature. The three letters IHS are often found on Catholic memorials but also occur elsewhere. The meaning of the letters can vary: they are the first three letters of Jesus in Greek, though they also serve as an abbreviation for 'In Hoc Signo' which, when accompanied by a cross may be a more likely meaning. They have also been explained as 'In His service', and other meanings may have been applied to these letters when established in use. This is an example

Figure 34. Catholic headstone with Cricifixion scene with altar and candles, and other features including cherubs, St Peter, and an hourglass. Monasterboice, Co. Louth, Ireland.

of a religious symbol which has been ascribed different meanings according to denomination, and also over time as theological emphases have altered.

The iconography associated with Roman Catholic memorials can also be readily identified. Images of the Virgin Mary, the sacred heart, saints, or the Pope frequently occur on memorials up to the present day. Other designs most popular on Roman Catholic memorials can also occur on others, however, and some caution is necessary if this is the only evidence. Such designs include the crucifixion, hands in prayer, or the lamb of God, which are all very common on Roman Catholic memorials but can also be found on those of other denominations.

A large scale survey in Louisiana revealed significant regional trends which also reflected denominational preferences. More Catholic memorials had crosses or sculpture, or other religious symbols, and there was far less concern with an orientation with feet to the east. (Nakagawa, 1994).

1.1.2 Orthodox

The Orthodox church traditions are well represented not only in the eastern European homelands where little work has been undertaken, but also in western Europe and colonial situations (Graves, 1993). They have, however, been mainly considered within an ethnic context. The purely religious aspects of the symbolism, and changes over time and space, have yet to be explored.

1.1.3 Protestant

The Protestant tradition can be seen in the avoidance of phrases used in the Catholic memorials, since there is perceived to be no requirement for prayer for souls as redemption is based on faith (and in some sects, also actions in life or membership of an elect group). There are, though, fewer direct positive indications. This was noted in Louisiana where most of the positive correlations were with Catholics, with the exception of orientation with feet to the east which was a widespread Protestant feature, and applied to all such congregations (Nakagawa, 1994). Within the Protestant tradition, the various denominations can in some respects be seen to operate with different priorities and ideas of organization, though this may as much be related to their time and place of development as their theological stands. The ways in which various Protestant denominations may be seen to differentiate themselves has been considered at a regional level in north Pembrokeshire, Wales (Mytum, 2002e).

One group which can be seen to follow elements of their beliefs through into the burial record is that of the Society of Friends, the Quakers (Stock, 1998a; Stone, 1991). Burial grounds of the Quakers are typically filled with small, similar monuments which are simple in shape and lacking decoration. They state the name of the deceased and date of death, with either date of birth or age. Months are given in numbers, as the names of the months were derived from pagan sources and so were not to be used by Friends. In some burial grounds the similarity of the memorials is very strong, with limited changes only over time. In other cases there was a little more variety and the rules were not enforced to the same extent. Recent excavation of Quaker burial grounds in southern England has shown how the rules regarding the form of grave and nature of burial were less strongly enforced (Stock, 1998b). The textual evidence lays great stress on the simplicity of burial, though from the evidence of debates there were clearly differences of opinion within the Society of Friends over this matter (Stock, 1998a). The excavated evidence shows that sometimes elaborate coffins could be used and not all graves were simple earth dug features but some were lined with stone or brick, and could be structures for single or multiple interments. All the elaborations were certainly not within the written rules.

The Shakers developed from the Quakers and had much the same attitude to memorialization. From the small numbers of early stones that survive, they were more successful in enforcing conformity, though the later replacement of memorials with others of standard form, or even a single communal memorial, has unfortunately reduced the amount of original data (Malloy and Malloy, 1992). Indeed, it is possible that the later changes may have removed any earlier variability. Mennonite memorials in three cemeteries in Headland County, Ontario in the period 1870–1899 were uniform in height, but did display considerable use of religious decoration (Norris, 1988:142–43).

1.2 Other Religions

1.2.1 Mormons

Mormon attitudes to death have an impact on both death rituals, mourning and general attitudes to death (Meyers, 1975). Some combinations of symbols seem to recur on Mormon tombstones, and these have been the subject of several studies (Cunningham, 1989; Schoemaker, 1989; 1992). Early Mormon gravestones do not seem to have been so distinctive, however, and the masons supplying the first generations in Salt Lake City used the repertoire of designs and generic motifs which they has acquired in their homelands (Edison, 1988). During the 20th century, representations of the Mormon temple have become more frequent on North American memorials (Thursby, 2003).

1.2.2 Jews

In central Europe there are numerous Jewish burial grounds in the countryside, and a few in cities such as Prague (figure 35), with memorials in some numbers dating back to the 15th century (Sadek and Šedinová, 1989). There has

Figure 35. Jewish cemetery, Prague, Czech Republic.

been a limited amount of research on these which has focused on identifying such sites and considering some of the artistic features of the monuments (Ehl et al., 1991). There are also extensive current programs of recording, though how far this will go beyond the genealogical is unclear. Bronner (2002) has discussed the legend of how the Remu cemetery, Cracow, Poland, was not completely destroyed during the Nazi occupation, linking graveyard studies to folklore.

Jewish memorials show a great diversity within the New World (Halporn, 1993). The variation in memorial type and design has been examined by Gradwohl (1993a; 1993b). This has highlighted the complexities of variation within Jewish Reformed, Orthodox and Conservative communities. The Sephardic tradition has been the subject of a separate study (Gradwohl, 1998).

Jewish burial grounds were often placed apart from Christian ones, though they may be in close proximity or even incorporated within a cemetery, particularly when burial areas were being reorganized in the 19th century on the edges of expanding towns. For example, in Gibraltar the Jewish cemetery was adjacent to but completely separate from the North Front Cemetery used by the rest of the population of the British colony (Mytum, 1993). In contrast at Mobile, a Jewish section was established within the Magnolia cemetery. Once filled, however, separate cemeteries for the Reformed and Orthodox (now Conservative) congregations were established on nearby blocks (Sledge, 2002).

1.2.3 Other Non-Christian Groups

Other religions provide places for burial where necessary, though the Hindus use cremation and provide no permanent memorials. Moslems have their own tradition of markers, including vertical pillars of wood or stone, elaborately carved and indicating by shape and decoration features of that person. There has been little research on such memorials. Some Moslems settled in western countries have monuments of a type typical in that context, but with their own text and symbols on the monument.

Chinese burials can now be found in many major cemeteries, and when not influenced by Christianity, can reflect traditional beliefs involving Buddhist, Confucianist and Taoist traditions (Blake, 1993:85). A study of Hawaiian Chinese burials and monuments highlights the process of burial, excarnation to collect the bones and their return for burial in China (Purnell, 1993). Those that were excavated but not repatriated were placed in vessels and placed in the "Bone House". The Chinese purchased land for their own cemetery in 1851, on a location selected according to Feng-shui (Purnell, 1993). A study of Chinese burial monuments in Hong Kong provides a comparison based on Chinese in their homeland (Chow and Teather, 1998).

Japanese burials occur particularly on the west coast of America and Hawaaii. First arriving as plantation workers in the late 19th century, the Hawaiian workers had to be buried despite being Buddhist and wishing cremation, as there were no facilities; they have simple burial markers with inscriptions. Cremation facilities were in time made available, and in the 20th century a range of traditional Japanese *hakka*, graves or grave houses, are used. Most Japanese cemeteries have become filled and burials now take place in Buddhist sections of cemeteries. The later Japanese memorials are often in black or gray granite and may be inscribed on both faces of the headstone. A few Pagoda markers are known which have five parts indicating the elements air, wind, fire, water and earth (Purnell, 1993).

1.3 Combined Burial Grounds

1.3.1 Churchyards

In most cases burial grounds associated with a place of worship, whether adjacent or placed at some distance, are for those of that particular religious affiliation alone. This is not the case, however, for those belonging to what are termed the established churches still in England and previously in North America, Wales and Ireland, which acted as the state church and so the potential provider for everyone. In such as situation, any inhabitant of a parish normally has the right of burial in the local churchyard. As many denominations were only permitted their own burial grounds in Britain and Europe in the 19th century (the income from burial fees was an important source of revenue for established clergy), the graveyards may contain those of many denominations. Sometimes those of this different persuasion were keen to emphasize this through the inscription or type of monument chosen; even where not, one has to be aware that the population of memorials does not necessarily reflect the dominant ideology of the controllers of that burial ground. Some separate burial grounds were allowed from the 18th century, and became common during the 19th century. From this time most burials that continued in the established churchyards were of that denominion unless a family wished to continue to use its burial rights over family graves or vaults. The parish graveyard remained open to anyone, and as a result small religious minorities continue to use the burial grounds of the established churches and may even have small zones set aside for them.

1.3.2 Cemeteries

With the development of cemeteries, different zones of the cemetery were set aside for the various groups of denominations. Their size and location may have reflected their importance at the time of the cemetery's foundation, and some denominations may have spread into adjacent areas over time, or had extensions built specifically for one denomination. Much could be done by examining the spatial dynamics according to denomination over time, and the ways in which such differences were highlighted or hidden in the repertoire of memorial forms and decoration placed in such zones.

2. ETHNICITY

Ethnicity can be a powerful factor in social definition, and has been a popular subject of study with regard to mortuary behavior, and this has been recognized within the literature on memorials (Meyer, 1989a; 1993a; 1993b).

Ethnicity is self-defined rather than inherited (Jones, 1997), and many aspects of mortuary behavior in a group's homeland will not necessarily carry an overt cultural association. Funerary customs, burial and commemoration will be part of normative culture, but when immigration or emigration highlights differences between groups, then such behaviors may become powerfully overt. They may change under processes of adaptation to new circumstances or acculturation, and in that respect mortuary behavior is no more resilient to change than, many other aspects of culture, but it can still retain potent associations. Although European countries have experienced considerable mobility of populations, most archaeological studies of ethnicity have so far been carried out in colonial contexts. Memorials have proved a particularly popular area of investigation. For example, Norris (1988, 142–144) considered the difference between Italian, British and eastern European burial markers in one cemetery in Welland, Ontario over the period 1890 to 1919 and found significant differences, and Stone (1991) noted ethnicity as also a factor in Long Island memorials. Meyer (1993a) provides an extensive annotated bibliography for American ethnic diversity as studied through memorials.

The Irish diaspora offers opportunities for examining transfer of cultural traits to North America and Australia. The shamrock is noted as an indicator of Irish origin in the Parramatta cemeteries, though no analysis of this data would seem yet to have been undertaken (Dunn, 1988; 1991). Preliminary survey in Ireland by the author suggests that the shamrock, Celtic cross motifs and the use of Gaelic on memorials develops in some parts of Ireland during the later 19th century; some of these strands relate to the Celtic revival in a range of artistic media. The freestanding Celtic cross became a widespread though relatively expensive memorial type, and is found across the world in Britain, America and Australia. Whilst often chosen by emigré Irish, its very success as a memorial form meant that it was not always being used as an ethnic symbol, and just became part of the repertoire of revival sculpture alongside obelisks and sarcophagi. The Scotch Irish on the Carolina Piedmont took their northern Irish armorial memorial tradition, and modified it to produce unofficial heraldry with more American symbolism (Clark, 1992).

Protestant German settlers can be identified not only through language but also the choice of iconography in 18th-century Pennsylvania (Graves, 1988, Farber and Farber, 1988). The art on the memorials can be linked to motifs occurring in other aspects of Pennsylvania German culture. They include motifs such as the tree of life or a flower such as the tulips to represent it, hearts, and a range of astronomical features including the waning moon as well as the sun and stars. A significant number of stones have no inscriptions but are thought to be early because of their form and iconography, and it is likely that the texts were painted onto the blank panels of the stones (Graves, 1988:74–76). Victorian stones show greater acculturation in terms of motifs, but the style and content of inscriptions, with full birth as well as death dates, continued. Metal crosses are also a feature of German settlers in America (Horton, 1997; Jordan, 1982; Kloberdanz, 1983) and also in Canada (Patterson, 1976), though many are also used by other central European groups such as the Czechs (Anderson, 1993; Horton, 1997). Other ethnic groups which have been the subject of several studies include Czechs (Kiest, 1993), Gypsies (Erwin, 1993), and Italian-Americans (McGrath, 1987; Matturri, 1993; Inguanti, 2000). The Navajo, Mormon and Zuni funerary and commemorative traditions of New Mexico have also been recorded (Cunningham, 1989; 1993).

Studies of African-American burial culture, as opposed to markers erected by whites for slaves or ex-slaves (see below), have often been concerned with identifying relics of African belief systems and practices, and the development of ideologies and behaviors which relate to the American context (Blessingame, 1972; Vlach, 1978). Much of the data is relatively recent, however, and patterns of continuity, rediscovery and development of new traditions are all possible. Antebellum material culture studies have indicated the rich and varied repertoire of items used in the construction of memorials. Later 20th century memorials from North Carolina have been surveyed and the importance of concrete markers and plot enclosures was noted; there is also a significant number of unique, home-made memorials in various materials, often reusing other items (Little, 1989; 1998; Nigh, 1997).

Mexican-American cemeteries have been the subject of a number of studies in southern California, New Mexico and Texas (Barber, 1993; Jordan, 1982; Sanborn, 1989; Stilgoe, 1978; Warren, 1987). Early memorials were probably simple wooden markers in the form of crosses, which are replaced later by concrete crosses, sometimes quite elaborate in design. They and other monument forms can have niches for holy objects or pictures, or an item relating to the deceased. Graves may be defined concrete curb, or a more visible *cirquita*, an upstanding grave surround.

Acculturation processes have been noted in a number of studies, including Chinese burials in America (Blake, 1993), and the Czechs and Moravians of Texas (Eckert, 2002). The conversion from wood to stone in the case of Tsimshian Indian memorials of western Canada in the late 19th and early 20th centuries has been studied by Hawker (1990; 1991); recent American Indian memorials have been examined by Gradwohl (1997). Australian Aborigine burials have been considered in a mission context (Wrightson, 1998).

Cemeteries for those who died abroad, such as the Protestant cemeteries in Rome (Nylander et al., 1989; Rahtz, 1987) and Florence (Freeman, 1993), present their own opportunities. The overall cemetery design and the memorials within tend to emphasize ethnicity over religion, perhaps because their very location indicates belief. This can also be seen with regard to many of the memorials in the Indian subcontinent (Harfield, 1984; 1985; Radford and Farrington, 2001; Yalland, 1985), where despite some use of indigenous motifs, most memorials reflect Classical revival styles which emphasized European cultural traditions. Here, many monuments include crosses and other Christian iconography because the indigenous religion was not Christian, unlike those in the European context where some of the ideology could be taken for granted.

3. LINGUISTIC GROUP

Most memorials are provided with the inscriptions in the single language of that cultural group, and this can itself be an important indicator of affiliation in a shared burial ground such as a large city cemetery. The use of more than one language may be to indicate scholarship (as with the incorporation of Latin or Greek in memorials) but often suggests a multi-lingual lifestyle. The weight given to each language, and what aspects of the inscription are in which, can be important indicators of acculturation and conscious manipulation of the medium of the memorial. After all, only certain parts of the text may be readable to some of those who pass by the grave. The influence of one language on another in grammatical and syntactical ways can also be considered, as can the format and content of the inscription. Theoretical studies involving sociolinguistics have considered language use in German graveyards and the Taylor City cemetery, all in Texas (Baird, 1992; 1996). The Chinese texts at Valhala cemetery, St Louis, Missouri, have been analyzed to discern changes in language use and ability to write text, and to identify various Chinese dialects suggesting the degree of heterogeneity and origins of the community in the city (Blake, 1993). A similar interest in changes in linguistic abilities and aspirations over time has been undertaken on Czech memorials in Texas (Eckert, 1993; 1998; 2001).

Where groups are bilingual, the choices regarding language can be more complex (figure 36). Trends in acculturation, transculturation, domination and resistance can all be noted. The ability to look at such trends over time is especially valuable. In cases where the population has been fixed but language change has been encouraged from outside can be illustrated by research in Wales (Mytum, 1994; 1999b; 2002e). The alternative situation is where a substantial number of immigrants arrive, using their native tongue within an environment dominated by another language. Here the choices of the first and second generations may be compared, with subsequent generations reflecting further integration or continued

Figure 36. Left, Headstone with Chinese and English, Bournemouth, Dorset, England. Right, Headstone with Biblical texts in Irish, Latin and then English, Monasterboice, Co. Louth, Ireland.

cultural identity by retaining the use of the language in some ways on the memorials. The process by which the Pennsylania German memorials become bilingual has been examined by Graves (1988, 82–92), and the role of carvers in providing stones in German or English is explored by Veit (2000).

4. VOCATIONAL IDENTITY

Males in particular are often identified by their vocation. This most often occurs through the text, but this can be replaced or augmented by symbols or pictorial representations (figures 28, 37). The textual forms can be with a single word or line of text after the name, or can be through the use of titles which indicate a profession. Most frequent are medical and judicial responsibilities, though political and religious vocations are also frequently found. Military titles are discussed later. Certain trades tend to be better represented than others and give more detail, so master mariners often also name the type of ship and its name, perhaps also giving

Figure 37. Memorial to an artist, Riga, Estonia.

its port of registration. Government officials may also give details of their role, and perhaps several posts that were held during their career.

Sometimes a scene of activity may be included, though this occurs more often in 19th-century and later examples of this style of identification. Mariners may depict their sailing vessel, farmers a plowing scene, with such designs produced incised or in low relief. With the recent developments of etched designs on polished granites a much wider array of designs are possible (figure 28), now even taken from photographs and reproduced with the aid of computers. The loggers of the American north west have been studied with regard to images and the selection of epitaphs and textual descriptors to show that a particular cultural identity was and still is manifest through grave markers (Meyer, 1994).

An emphasis on trade or profession is not universal, being absent completely in some areas and rare in most. In some regions at certain times, however, this is a dominant feature, though even here not of all vocations. Thus, in 18th entury Yorkshire occupation is rarely displayed pictorially, in lowland Scotland and parts of Ireland at the same time the use of trade symbols is widespread. Tools of the trade may be depicted, or the coat of arms of the trade guild may be used (McCormick, 1976; Willsher and Hunter, 1978). This follows on from a tradition on medieval grave slabs where one or two features such as a sword for a knight, or a crosier for a bishop were used. Some later examples are as simple, though many became much more elaborate. The stonecutter Louis Brusa designed his own memorial, which was made in the year if his death, 1937, by the sculptor Doanto Coleti. Killed by the effect of granite dust, he is depicted slumped against a rough slab, consoled by his wife. Whilst the sculpture is in a contemporary realist style, it has resonances with earlier European artistic traditions (McGrath, 1987).

Where working place burial clubs operated to support families in the costs of the funeral and interment, it is possible to find burial grounds established or sections of cemeteries purchased for use by such groups, thus creating a zone based on trade (Florence, 1997). This degree of spatial segregation is rare, however, and was of course not exclusive since some relations entitled to burial through the club would be of different occupations. Spatial segregation by occupation has been noted as a feature of the St Andrew's Ukrainian cemetery near New Brunswick, NJ, though where both spouses worked in different occupations it was the husband's that dictated grave location and both would be buried there (Graves, 1993:55).

4.1 Religious Leaders

The memorials of religious leaders were often important monuments at the time of their erection, as they reflected not only a family's loss, but that of a community and wider organization. Their memorialization could be differentiated in several important ways from their flock. The portraits of ministers, and the use of table tombs, have been considered for Congregational ministers in part of Massachusetts (Malloy and Malloy, 1997).

With regard to orientation and location, this can be clearly seen as important in Roman Catholic burial grounds where in death as in life, priests are often placed together in a separate area. This emphasizes both their difference through their vocation from the rest of the populace, and also their corporate identity, being placed together, yet commemorated by single, separate memorials marking the individual inhumations. The style of memorials is often also distinctive compared with the rest of the graves at that burial ground, though only rarely of a type never used within the wider population. Besides references in the inscription, the use of the chalice is often found to emphasis the crucial and unique role of the priest in mediating between God and mankind.

In Anglican churchyards, the memorials of the clergy are notable in that they usually face the opposite way to the other memorials. On the Day of Judgement when the dead will rise to life, the clergy have to face their flock to prepare to lead them into the life to come. They, too, are sometimes indicated by the chalice, or a more widespread use of the cross motif when it is not used by others. Clergy may also be buried within the church, normally in the chancel, even when intramural burial for almost all others had been prohibited. This again emphasizes the special role of the clergy, buried in that part of the church with vicar's stall, altar and communion rail, which was at the center of their role within worship.

Major church leaders, for example, bishops in the Anglican or Catholic churches, were commemorated in places of worship such as cathedrals, and could have more than one memorial—one where they were interred and another in the place where they had served an important part of their ministry. Many of the lesser churchmen who served particular communities, often for much of their lives, were purchased memorials by their congregations. This is often stated in the inscription and explains why some of the monuments can be so large. It must be remembered, however, than some churchmen in the Anglican church could be extremely wealthy in their own right, or through the incomes received from the church benefices, so they could have afforded substantial monuments without any external support.

4.2 Military

Military burials are often distinctive in some way. When death has been in action, the sense of loss is immediate and sudden, though not always unexpected. Many soldiers who have died in such a way, or through some accident, are commemorated not by their family but by their comrades in arms. Given the costs of such memorialization it is normally officers that were recorded in such ways, though the 'men' may also be given an honorable though not so often an individualized commemoration. Such memorials emphasize certain features considered desirable in an officer, and act as reinforcement of military conduct at a time when it is possible that some might question its rationale. Moreover, such memorials later serve as exemplars for future cohorts, and indeed the care with which such memorials are maintained and indeed replaced by the various services of the military up to the present day emphasizes the role of tradition and remembrance within the forces' ethos. Retired military figures are often commemorated with details of their rank, and even their awards. Military insignia can feature on the memorials also, emphasizing that link even at the expense of familial associations. War cemetery memorials have been discussed in detail in Chapter 4.

Communal memorials, first erected in numbers in the 19th century, but becoming a regular feature of local communities following World War I, are also of considerable interest (Borg, 1991; Inglis, 1992a; 1992b; 1998; Moriarty, 1995; Winter, 1995). These are truly memorials, not located near the remains of those listed upon them. They acted as foci for communal grief and still serve as such. For example, in Britain on Remembrance Sunday in November, those not only of World War I but subsequent conflicts are remembered by their families of the deceased and also the community as a whole (figure 38, left). Whilst most memorials of this type are community memorials, in villages, towns and cities, some are placed at centers of power reflecting the nation's loss, as with the Vietnam memorial in Washington DC (Hubbard, 1984). There are also regimental memorials and those erected by employers, such as those by the railway companies.

Figure 38. Left, Village World War I memorial constructed as the churchard gateway, Gilling East, North Yorkshire, England. Right, Woodmen of the World memorial, La Fayette cemetery, New Orleans, Louisiana.

It is the sad fact of war that many are killed whose bodies are never recovered, and there are those who are recovered but not identifiable. The monument to the unknown soldier is a phenomenon of the 20th century, allowing a focus of grief for those who have in all senses lost their loved ones and emphasizing to the community this even greater sacrifice by the living as well as the dead. Such memorials have been the focus of some recent research (Ingersoll and Nickell, 1987; Inglis, 1993).

The choice of symbolism, design, and location of communal monuments is one of special interest. As large scale features to be placed within the fanric of the community, there was great concern and resources applied to the design, with notable architects and artists involved. The written and symbolic statements of time, place and cultural association are even more overt in such monuments, and may be well have been at least partially documented in the historical record. Whilst always of their time, the memorials may be contemporary, look forward to a greater future won, or backwards to a cultural identity retained or regained. The deconstructing of policy and purpose, power and meaning, can be usefully followed through in the case of communal war memorials more easily than in many other monuments where individuals rather than committees were involved, as the latter produced minutes, financial accounts, and written records of many other kinds.

4.3 Other Identities

4.3.1 Enslavement

Whilst most slaves were not commemorated with a permanent memorial, a small proportion were. Some plantation cemeteries contain burial mounds for a small number of the interments (Handler and Lange, 1978), and others may have had some form of marker, erected by the friends and relations of the deceased. A small number of slaves were given memorials by their owners. These stones can be examined to consider the motivations and categorizations utilized by those who erected the monuments, and give us one perspective on the social identity of slaves (Garman, 1994; Krüger-Kahloula, 1989; 1993; Malloy and Malloy, 1994; Tashjian and Tashjian, 1989). Some memorials were erected by or for emancipated slaves, though again often it is unclear who erected them, and when.

4.3.2 Place

In many areas, people are defined by their place of abode; the familial implications of this have already been outlined, but other identities can be drawn from such information. It can indicate to contemporaries much about a person, since the street in a town carries with it a particular level of social standing, and a farm in the country would be known for its scale and relative worth. Many people, moreover, worked from their homes, and so to those reading the memorial their trade would be indicated by the address.

Place can be defined in inscriptions at a number of levels, and may be recorded in some detail. Whilst some may mention exact properties such as a farm name and even a house number in a street, many memorials are less precise but still significant. Many people are defined by parish or region. Place of birth can be stated, showing an identity with another region, and where place of death is of some distance away this may be implying a different form of relationship to the community within which the memorial is erected. Some memorials give considerable significance to the location of the actual body if not marked by the monument itself.

4.3.3 Society Membership

Membership of societies can also be a significant element in social definition (Gabel, 1994). The most commonly noted example is that of the Masons, whose symbols are often portrayed on monuments ranging from simple headstones to complex mausolea, as well as in other decorative art (Zaniello, 1980). Whilst the portrayal of hands on their own cannot suggest a Masonic connection as their distinctive handshake looks like any other, when found with other symbols it may be assumed that is what is represented (Patterson, 1989, 181). In contrast, membership is not so frequently mentioned in the text of inscriptions. Other societies are also found portrayed on memorials (Gabel, 1994), including the various Woodmen organizations (figure 38, right) (Roberts, 1990:190). The Modern Woodmen of America (founded 1883) and the Woodmen of the World (founded 1890) both offered insurance schemes and suggested marking the burials of members with tree stones (Ridlen,1996; Stott, 2003). However, the larger limestone memorials from Indiana do not seem to have been used only for this purpose and smaller granite memorials often tended to be selected (Roberts, 1990, 190).

4.3.4 Achievement

A small but notable number of memorials emphasize a particular achievement of the deceased. In most cases this is done through a combination of illustration depicting the event and text elaborating on it. In some cases it can be a feature which resonates with the activities to be recalled, such as the 19th-century explorer Sir Richard Burton's mausoleum in the shape of his field tent to be seen in Mortlake cemetery, London (Greenwood, 1990:310–12). Notable achievers may also be buried or remembered in particular places, such as great British cultural, military and political figures in Westminster Abbey in London. Epitaphs often recount the achievements of the deceased, though some are remarkably formulaic and may not indicate any specific qualities of the individuals being eulogized. Those who died young may have their academic or sporting achievements noted.

4.3.5 Manner of Death

Another significant identifier for a small minority is the unusual manner of their death. The trauma of the unexpected event leads the surviving relatives to wish this to be featured on the memorial; communal memorials may also be erected to those who suffered death together, such as in a boating accident. Though they may have nothing else in common other than their moment and manner of death, they may be buried and commemorated together. With accidents at work, the employer may pay for the memorial, either to an individual or to a group, the latter in some respects operating in the same way as a war memorial. Tragedies also often spawn other commemorative material, whether it be a 19th-century colliery disaster or the death of a head of state. Civilian casualties in war may also be specifically noted, as with those who suffered from the Lawrence massacre, in Kansas during the Civil War (Thies, 2001).

Figure 39. Headstone erected at the site of a road accident, Balrothery, Co. Louth, Ireland.

A recent form of memorial which reflects the manner of death is that of the roadside memorial. Many such features are transitory, though some become permanent, and are erected on the roadside at the point where the fatal accident took place (figure 39). Such memorials can rarely be read by those travelling along the road, except at risk of suffering a similar fate to the person commemorated, but they mark tragic loss and act, rather like the warning verses on 18th-century memorials, as reminders of our mortality and the fragility of life. They may therefore provide solace for a wasted life in terms of preventing other such accidents, though they may serve more of a political purpose in the demand for an improved and safer road.

5. CONCLUSIONS

Memorials, almost by definition, are making claims about the identity of the deceased. As every individual has many types of identity, those chosen reflect sometimes the wishes of the deceased but usually those of the close family. The selection and emphasis of competing and contrasting identities makes the analysis of memorials particularly powerful. This has been one of the aspects of mortuary research which has had sustained research, particularly on ethnicity, though there is still enormous potential for further study.

Chapter 8

Attitudes toward Death, the Body and Remembrance

Attitudes toward death can be explored in a number of ways using the material culture of death and combining it with documentary sources, oral history and ethnographic study. Various strands may be teased out on this subject, and they can be understood in part though examination of the normative patterns, but also highlighted through the eccentric and atypical.

The first strand is the attitude toward death and the afterlife of the bereaved. This can be considered through body treatment and aspects of the funeral, through burial itself, and any subsequent memorialization. This can be expanded by examining longer-term attitudes toward the grave that may be individual, or relate to multi-generational commitments to the burial ground. This can be gained from study of letters and diaries (Jalland, 1996) and illustrations and photographs of graveyards in the past, or from anthropological studies for the recent past and present (Bradbury, 1999; Hockey et al., 2001). A theme of recent concern has been the relationship between material culture, emotion and memory (Hallam and Hockey, 2001; Tarlow, 1999).

The second strand is to consider how society at large and those involved with the funerary crafts or professions treated human remains and burial grounds in general. This is relevant when considering how burials are grouped, to what extent earlier burials were disturbed, and the ways in which carvers rather than the purchasers affected designs on memorials. One interesting aspect that has only recently begun to be considered is what elements of the body preparation and funeral was seen and experienced by the family, or by the wider population, what was hidden from view and carried out by the funerary professionals. Present day attitudes to the dead, and their implications for archaeologists and anthropologists in conducting fieldwork, are briefly considered in Chapter 9. Here, consideration of the manner of death, the treatment of the body, and the ways in which remembrance took place are reviewed.

1. THE MANNER OF DEATH

1.1 The Good Death

Emphasizing the good death could be comfort to the bereaved and a lesson to other observers (Beier, 1989; Gittings, 1999; Houlbrooke, 1998; Llewellyn, 1991; Bradbury, 1999). This was a major concern in life and greatly affected the manner and behavior of those who were ill, as they prepared for their demise, particularly up to the 19th century, and for Roman Catholics the administration of the last rites was particularly important. The many warning epitaphs of the 18th and early 19th centuries are concerned with ensuring that others will be prepared for and have a good death. Many sentimental epitaphs of the 18th and 19th centuries make mention of long illnesses bravely borne.

Numerous 19th and 20th century inscriptions emphasize sleep and peace. What suffering these may hide from public gaze cannot be known, but the sentiments certainly pervade many memorials of this date in most regions and across classes. It may be that the memorial, designed to be a comfort to those left behind, tends to present a positive view of the process of most deaths, so that emphasis can be on remembrance of the person during all of their life, not during just the final phase.

1.2 The Bad Death

The sudden, tragic death affected family and friends more profoundly than an expected death from old age, and this is often reflected from the later 18th century onwards in the provision of more elaborate memorials in some such cases. The cause of sudden death may be mentioned on the memorial, and indeed can be the focus of decoration such as a shipwreck. Accidents involving many people may have communal memorials, which may or may not be where any burials took place. A wide range of disasters, from accidents in the construction and mining industries, to sporting accidents and terrorist incidents, can all have large scale communal monuments. Paid for by an employer or public subscription, they reflect a sense of communal loss felt by the wider community. These may also be augmented by personal monuments, where individuals have their own memorials or are added onto already existing family ones.

The one form of tragic death which could create a more adverse reaction was that of plague such as cholera. Both because of the scale of sudden burial, and the threats to the health of the living which the plague offered (though the causes were rarely understood in today's terms), interment and commemoration could take a different form. Many parish and community burial grounds could not cope with the sudden rise in demand, and separate, temporary, burial grounds were often established (Curl, 1972; Sloane, 1991). These may subsequently be marked with individual or communal memorials, though others have been covered by later developments.

The most obviously commemorated form of tragic death is that of the casualty of war, an offering for the nation which has been celebrated with governmental support and maintained with money from the tax payers who have subsequently benefited from the sacrifices made by those who died in past conflicts. As such, the memorials may be highly political, and their treatment can reflect attitudes to the conflicts in which the deaths occurred, such as southern Irish deaths in World War 1 (Taafe, 1999). Often, however, the war memorial stands apart from the politics, and is seen a loss for community rather than faction. This can be seen with war cemeteries containing victims from both sides of a conflict (though often buried in separate zones).

2. ATTITUDES TOWARD THE BODY

Recently, archaeologists, anthropologists and other social scientists have become interested in the ways in which the body was regarded, but surprisingly little of the theories generated by this work have been applied toward mortuary behavior (Arnold and Wicker, 2001; Tarlow, 1999a). Yet attitudes toward the body, both immediately after death and following its burial, can be examined through the evidence of the memorials and also by excavated evidence. The rituals of visiting the family and body, of sitting with the corpse, of attending the funeral and visiting the grave were extremely varied over time, space, social class and religious persuasion.

Changes in attitude to the dead body followed a pattern in many parts of the western world after the mid-18th century in which the corpse became seen as a spreader of death (Ariès, 1974). The later 18th-century Enlightenment, and disillusionment with formal religion, provided a context for supporting the scientific sanitary argument for new burial areas where overburial would no longer take place (though exhumation of clean, skeletal material to reuse space was accepted in much of Europe). France largely led this change (McManners, 1981), but widespread shifts in popular attitudes to death across Europe reveals similar trends. The aristocracy could escape church burial with family mausolea on their estates, first created in England at Castle Howard and then developed elsewhere (Colvin, 1991). Underlying all these changes was a shift in attitudes regarding the relationship between living and dead. Rather than emphasizing the decaying bodily remains, the successes in life of the deceased should be celebrated and remembered (McManners, 1981). The focus on the actual body was replaced with memories of the person alive, a clean, positive image not tainted by putrefaction. From this developed in the 19th century the 'beautification of death' which was a particularly strong trend in America (Bell, 1990; Farrell, 1980; Little et. al., 1992), though present also elsewhere (Morley 1971). It is important to remember that during the 19th century the image of heaven became more domesticated, with the family reunited, a theme that is notable in popular literature, personal correspondence (Douglas, 1977; Jalland, 1996; Morley, 1971), as well as on memorial epitaphs with phrases such as 'We shall meet again' and 'Reunited'.

2.1 Interment

Excavated evidence can contribute greatly to our understanding of both the process of interment, and attitudes to the repercussions of this form of body disposal. Written sources can also be valuable, as can images of funerals and interments.

2.1.1 Disarticulation and Its Avoidance

The most frequently encountered engagement with a burial occurs when another grave is being dug. Throughout the medieval period, bodies were cut through by others and the disturbed remains were either piled in a corner of the new grave or were placed in a communal ossuary. Examples of such ossuaries still survive and may be available for investigation (Roberts 1984). Many smaller fragments, or those already disarticulated, were just left within the soil, and were reburied when the grave was back filled (figure 40). This is still the behavior in ancient burial grounds in Britain and Ireland, where fragments of disarticulated human remains can usually be seen on the surface of new graves or in flower beds within the graveyard. In most excavations of ancient churchyards in Britain, only a small percentage of burials will be even partially articulated, and about 50% of bone by volume may now be completely dispersed and often broken.

In countries where human skeletal material is gathered into ossuaries, these repositories can have special significance. In Naples, Italy, prayers may be said in the charnel houses beneath churches, linking the living and communal dead (Ariès, 1985:167, 170).

Awareness of the fate of earth burials led to the elite's selection of family vaults, and the emerging urban middle classes use of parish crypts, though this did not necessarily lead to a quiet repose indefinitely, as discussed below. Another solution favored by the middle classes was to construct a brick or stone-lined shaft into which several interments could be made, separated by iron rods (Cox and Stock, 1995; Litten, 1985). A cheaper version was the single, shallower brick-lined grave (Litten, 1985). These were constructed mainly within churches (figure 5, left), but were also used by some in the churchyard to ensure long-term protection of the interment.

One of the main functions of the memorial is usually to mark the place of burial, and protect it from unintended disturbance. The very form of many

Figure 40. Section of graveyard fill with disarticulated skeletal human remains including skulls, representing many inter-cutting graves over several centuries. At the base of the section, the angular outline of slightly deeper grave cuts can be seen in contrast to the sand subsoil. Kellington, North Yorkshire, England.

memorials therefore served this very purpose. There is not necessarily direct symbolism for this, but rather the design of the monument—a ledger, chest or table tomb or headstone with footstone and perhaps a body or coffin stone protected the grave space (figure 41, left). The monument did not serve merely as a commemorative marker at which to grieve; the presence, intact and undisturbed, of the body was a major factor in the development of the monument. It is not surprising, therefore, that in the early development of modern burial markers that there is an emphasis on the burial plot and its ownership. Indeed, early stones in Scotland marked and commemorated the purchase of the plot rather than the memory of any individual (Harrison, 1990). The same can be seen with some of the early Irish memorials (Mytum, 2002b).

Current concerns with burial disturbance can be seen through much of the legislation and archaeological guidelines which reflect to a greater or lesser extent the wider cultural concerns over the disturbance of the dead (see Chapter 9). Building work can disturb human remains, and where burials are known beforehand the decisions made regarding their treatment can be informative. In America, where disturbance of remains is much more widely viewed with disgust than in Europe

Figure 41. Left, Headstone, gabled body stone and footstone, Holyhead, Anglesey, Wales. Right, burial plot with raised fround surface covered with artificial grass to hide the skeletal remains within the soil, La Fayette cemetery, New Orleans, Louisiana.

with its history of densely used burial grounds, there were some dramatic efforts to avoid disturbance. For example, when Center Church, New Haven, Connecticut, was to be rebuilt, the chosen location was that of the community's early burial ground. Rather than disturb the burials the church was raised up and the crypt which was built beneath not only protected the graves but also allowed for the markers to remain upright in their places (Williams, 1992).

Where disarticulation is inevitable, because of repeated burial in a small plot, this may be a cause for concern. Whereas in Ireland large and recognizable fragments of human bone may be left on the surface of newly dug graves, in America this practice is not generally seen as acceptable. Thus, in New Orleans plots where earth burial in raised soil beds takes place, artificial grass can be used to cover the surface so that the bone and coffin fragments cannot be seen (figure 41, right).

2.1.2 Post-depositional Movement

The other way in which bodies are disturbed during everyday maintenance of a burial ground is when interments are placed in a crypt or vault. Even in family vaults there is remarkably little concern for the oldest burials which are often pushed to one side and suffer the indignity of collapse (Litten, 1999; Mytum, 1988). They may be left in that state or are piled up in a corner. The anthropological implications of families being buried together, not only as distinct and articulated corpses besides each other but at a later stage as intermingled disarticulated bones have not been greatly explored in the British context, though well recognized in the Mediterranean (see below).

In communal crypts, such as those at Spitalfields, London, even quite recent coffins could be treated with limited respect. As the various vaults within the crypt filled up and space was limited, coffins were rearranged, piled high as the roof, and placed on end as necessary (Adams and Reeve, 1987; Reeve and Adams, 1993). This lack of regard for the arrangement of family groups or individuals must have been due to the fact that families did not enter the crypt or family vault but remained above ground in the church when the coffin was taken down. Thus only a small number of individuals, and ones who had become immune to the concerns of the mourners, were involved in the maintenance of these burial spaces. Their lack of concern shows how, with the relaxation of normative cultural values, or at least control over the enforcement of these values, behavior can rapidly change in response to problems such as lack of space.

2.1.3 Body Theft

Subsequent deliberate disturbance of individual earth burials was rare in Britain, though the danger of body snatching for medical purposes was considered a major threat in some parts of Britain, particularly in the later 18th century and up to 1832 with the passing of the Anatomy Act (Gittings, 1984; 1989). This led to the development of various physical methods to prevent removal of human remains from their repositories. Some graveyards were walled and guarded with gates and even sentry boxes, as at Warblington, Hampshire (Bailey, 1987:93–94). In other cases, morte-safes, or iron cages, were placed over the graves to prevent exhumation; these were rented for a time and then made available to protect a new inhumation. (Ritchie, 1911; 1920). Other protection measures included the placing of a large rock over the grave; at Pannal, Yorkshire, a stone weighing over one ton was rented out for two weeks at the cost of one guinea (Burgess, 1963:56). This 'Resurrection stone' is still in the churchyard, though many other such items have probably been removed from other burial grounds. For some, security came through the provision of a locking iron coffin, or the placing of iron bands or chain round the coffin, practices identified in the Spitalfields, London crypt (Cox, 1996:106–109; Reeve and Adams, 1993:81–82). Another tactic was to leave the body, in its coffin, in a locked repository so that it had decayed too far to be worth stealing. This necessarily separated the interment from the rest of the funeral ceremony. These concerns were mainly felt within easy reach of the medical schools in Britain, and the extent of this activity in other countries is uncertain. Linked to medical advances came the desire to carry out autopsies following some deaths, and archaeological evidence has been recovered to indicate its frequency and demonstrate the methods employed (Molleson and Cox, 1993: 87–89).

2.1.4 Ossuaries

Mediterranean traditions saw the body as part of a larger community, a continuation of Catholic medieval traditions. Here, the movement of bones to a family or settlement ossuary was and still is frequent (Danforth, 1982; Goody and Poppi, 1994; Mytum, 1993). This attitude to human remains is one that is very different from the of the English-speaking world, though manifest in parts of the southern United States of America due to cultural influences. The arrangement of bones in elaborate patterns within the French catacombs, used in the 18th century as a venue for parties and still a tourist attraction, illustrates the diversity in attitude to human remains (Ariès, 1974; 1981).

2.2 Cremation

Behavior after cremation offers a different insight into attitudes the death and the human body. These can be considered in a variety of contexts. Societies promoting cremation were established from the 1870s onward in a range of countries including Italy, Germany, the Low Countries, the Unites States of America and Australia (Jalland, 1999:249; Sloane, 1991:143–156; Zelinka, 1991:47). The same arguments used to support the establishment of cemeteries to replace churchyards were once again applied. However, whilst cemeteries increased the chance for material expression associated with mourning and commemoration in a more expansive setting, the cremation argument was in part fueled by the rising weariness with the pomp of the funeral (Leaney, 1989). Nevertheless, progress was slow and only a few British crematoria were constructed in the final decades of the 19th century, with legislation in Australia in 1891 and Britain in 1902 providing the legal framework. Nevertheless, it was only after World War II that it began to become popular in Britain (Jupp and Walter, 1999:265). Cremation was accepted by the Anglican church in 1944. Cremation began in the late 19th century in America, following pressure from various cremation societies, but there were few crematoria. The Vatican's prohibition of cremation for Catholics in 1886 further suppressed demand (Sloane, 1991:140–145); it was only allowed for Roman Catholics from 1965.

The rise of cremation was possibly due to an increasing repugnance of decay. Cremation avoided this completely, and in the American context this same problem was dealt with through embalming (Davies, 1996). This cultural shift away from the body and decay probably in part accounts with the fascination of scholars and the public with earlier use of embalming in ancient Egypt and elsewhere. Trauma and discomfort may be felt by investigators dealing with corpses in an archaeological context; this is seen as natural (Reeve and Adams, 1993; Thompson, 1998; White, 1998). It may, however, be a contemporary constructed element of our own culture which needs to be recognized here, though no less serious for that.

2.3 The Unburied Body

In some cases a body was not buried, and so the emotions of the bereaved cannot be focused as they can be with the typical formal funeral rituals. Individuals may have been lost at sea, or died abroad and, even if buried, their relatives could not experience the event. Many were nevertheless commemorated at home, and this was no doubt done as part of the grieving purpose. It also placed the missing person socially within family and community, and enabled the social relations of those left behind to be negotiated through familiar strategies.

The community war memorial serves as a focus for grief near the place of abode of the deceased. The most potent symbols of the military bad death is where the body is not recovered; the memorials to the Unknown Soldier hold a particular poignancy (Ingersoll and Nickell, 1987; Inglis, 1993). Some large scale communal disasters such as mine accidents, or natural disasters such as earthquakes did not always lead to recovery of the corpses. How people could be remembered through a material form in such contexts has not been greatly analyzed.

Many Chinese who died abroad wished for repatriation of their remains; this indicates a powerful link with the homeland that transcended newly established social networks (Blake, 1993; Purnell, 1993). Some western figures were also returned home for burial, though generally only those of great standing, such as Lord Nelson who was brought back to England from Trafalgar, via Gibraltar, embalmed in wine to survive the journey. Most, however, were not returned, and even the fate of the corpse may have been unknown. Thus, many individuals commemorated on memorials may in fact not be present below-ground at that location.

2.4 The Body through Representation

Whilst the skeleton used as a mortality symbol was not personalized, there could be representation of the specific body on a memorial. Internal monuments have had a long tradition of sculptural figures, including those who are commemorated (Ariès, 1985; Kemp, 1980; Llewellyn, 2000). Full sculptural figures are necessarily expensive, so have always been rare as external monuments. They have tended to be linked to either extravagance or extreme grief, often associated with the death of children (Brown, 1994). Many figures in this genre are depicted

Figure 42. Portrait headstone, Lexington, Massachusetts.

in full health, though often in a relatively static pose. There are, however, a number of deathbed scenes, most frequently associated with women who have died in or soon after childbirth, often depicted with their offspring (Brown, 1994:179).

Portraiture as a less expensive way of representing the body could have been more widely available, but remained remarkably limited in its use (figure 42). Portrait busts, often in a material other that of the rest of the memorial, were particularly popular, however, from the later 19th century onwards in parts of Continental Europe (figure 43, left). They tend to depict the deceased as a mature adult, more usually males than females. Representation through photography in contrast rapidly became popular for both men and women in many parts of Europe (figure 44), and thence to America (Brown, 1994; Ruby, 1995). In recent years various methods of reproducing photographic images onto granite gravestones have been developed, with the result that these are found in many parts of the world.

The choice of image, particularly in a photograph, relates to another important facet of representation. This concerns the beautification of death, in that the image used on the memorial may show the deceased in a particular context. In this way people can be fixed in time; their images reflect a certain stage of life, alive, active, and sometimes with other members of the family (Hallam and Hockey, 2001). Evidence of old age, illness or accident can be wiped away, the face looking out from the stone can in this way project the idealized life of the deceased, and the essence of that person as seen by the grieving relatives.

Figure 43. Left, Busts on the front of catacomb burials, Pistoia, Tuscany, Italy. Right, Back of a headstone with mortality symbols, Killeevan, Co.Monaghan, Ireland.

Figure 44. Plaque over *loculus* with photographs of the deceased, SaPobla, Majorca, Spain.

2.5 Symbolism and Epitaphs

2.5.1 Mortality Symbols

The variety of death symbols seen on gravestones and other monuments is truly large, derived from a wide range of cultures and periods, though not necessarily retaining their original significance. The funeral symbols are often found in association with the skeletal elements, thus emphasizing the process of decay (figure 43, right). This tradition reaches back to medieval cadaver monuments (Ariès, 1974; 1985), but is one which was also separately developed in different times and places through repeated independent experience of decaying corpses and the evidence of bones in burial grounds.

A very popular group of symbols are those associated with the funeral and interment itself. The elements most frequently portrayed are the sexton's tools to dig the grave—the pick and shovel—and the coffin itself. In Scotland and the north of Ireland the bell used to ring the death knell is also often included (Willsher and Hunter, 1978; Mytum, 2002d). Some memorials have all of these symbols, but often only one or two were used, sufficient to evoke an appropriate response in the observer. Some of the tools are carefully illustrated, and can be identified as specific regional or temporal types. The same can be seen for those coffins that are depicted in more than simple outline. The form of the coffin is normally the hexagonal single-break coffin, and on well executed designs it can be possible to identify other details. The arrangement and form of handles, the use and design of coffin pins which held on the fabric covering of the coffin, and the presence of plates for inscriptions or decorations on the coffin are sometimes illustrated. These are useful depictions to compare with other contemporary illustrations, and the few survivals of hand bells and parish coffins that still exist (Litten, 1991:146), and the excavated evidence of the coffins which is discussed below.

The use of shrouds rather than coffins was a common practice for many poorer sections of the population until during the 19th century, and in some regions was almost universal. The shroud is used as a symbol on some memorials, most notably some of the New England headstones where it is held ready by a pair of imps. These are often seen, for example, on Joseph Lamson stones where 26 such designs have been recorded (Tucker, 1993: 162–163); such a design also occurs, though less frequently, in Britain.

The depiction of the whole skeleton or particular bones on memorials to indicate the decay was common. Sometimes echoes of the earlier styles can be noted, with worms creeping through a skull's eye sockets. Skeletal remains may be depicted in a range of ways, from the skull, with or without the lower jaw, to single or crossed long bones, up to complete, articulated skeletons. The skeletons are rarely prone, but are standing or active, sometimes holding other symbols of death, and even throwing Death's arrow or dart. The style of depiction of human remains varies from region to region, and even according to carver, but whether highly realistic or extremely conventionalized, all could be recognized and so encouraged the observer to pay attention to the monument's message through the iconography.

Other symbols of death abound. The figure of Father Time, armed with his scythe, can be found on New England headstones (Ludwig, 1966: 88–100) and tombs in Gloucestershire, England, for example. He is depicted in varying degrees of detail, and may occur with skeletons (figure 45). Symbols of death may also be more abstract. The burning torch pointed down to extinguish the flame of life is a popular neo-classical motif, and the broken column likewise symbolizes a life cut short (though often, because of its expense, is erected to elderly successful men!). Motifs in a similar vein are the sheaf of wheat, a sickle cutting wheat, or a flower with a broken stem (Hobbs, 2002).

Tarlow, basing her views on a study of Orkney monuments, considers that the mortality symbols relate to a clear understanding of the separation of body and soul (1999a, b). The very representation of the physical decay of the body is only to highlight the salvation of the soul. It is noteworthy that from the internal memorials in Tarlow's study, and from the early external memorials at St. Andrews onwards (Fleming 1931), Scottish and indeed British memorials often have a mixture of mortality and other symbols (including cherubs) from an early

Figure 45. Headstone with Father Time with scythe and hourglass on head, and skeleton opposite. Note the winged cherub on the top of the stone, Salem, Massachusetts.

date (Willsher and Hunter 1978, passim). The same pattern of central cherubs and marginal mortality symbols can be seen on early stones in the English Midlands (Barley 1948). However, one might argue that the admonitions and warnings in the epitaphs make clear the uncertainty of salvation, and that there is more ambiguity and concern here that is worthy of further investigation. In Scotland and northern Ireland, the use of Adam and Eve scenes can be seen to follow the same purpose of warning about the fall from grace (Willsher 1992a; McCormick1976), perhaps to be redeemed through faith in the risen Jesus Christ. This more positive message is not always explicitly stated or represented iconographically.

The New England symbol that encapsulated physical death and eternal life of the spirit was the winged skull (Tashjian and Tashjian, 1974). Rather than the British practice of using different symbols to represent mortality and eternity, this New England solution was to combine these two strands in one image (figure 46). The treatment of the skull varied greatly over time and space, becoming more face-like in many cases, and so acting as a transition to the cherub motif. These changes have been linked to individual carvers' views or those of society as a whole (Benes, 1977; 1985; 1987; Deetz, 1977; Ludwig, 1966; Tashjian and Tashjian, 1974). The relative confidence in salvation can at times be discerned in the design and use of symbols, but it is difficult to know if this is merely the carver displaying iconographic knowledge, the deliberate fulfillment of the clients' wishes, or merely variation within an established repertoire. Much depends on one's assumptions

Figure 46. Headstone with finely carved winged skull, Marblehead, Massachusetts.

about the levels of symbolic understanding of the contemporary audience; this was probably very high in Puritan New England, but perhaps lower in the less theologically obsessive Anglican areas such as Virginia and Carolina. Benes (1977) has suggested that Puritan concerns about the uncertainties of salvation may have been a major factor in the development of many of the skull designs found both in New England and in parts of Kent, England where nonconformists were most numerous. Thus, the link with the denomination may be crucial. The evidence from Ireland suggests that the nonconformist Scotch Irish predilection for mortality symbols (McCormick, 1983) was also taken up by Catholics who erected their own headstones using such symbols (Mytum, 2002d). These may have had meanings and associations linked back to counter-Reformation ideas about the body and salvation, manifested in the use of actual human remains and carvings in displays seen in communal ossuaries (Etlin, 1984). Thus both Protestants and Catholics could use the mortality symbols with their own particular emphasis on sin and way to salvation.

2.5.2 Warning Epitaphs

Linked to many of the illustrations of skeletons and symbols of death are warnings from the grave. These epitaphs are written in a style where the deceased speaks to the visitor to the grave. The personal pronouns give a sense of relevance and indeed urgency. Many such verses are just four lines long, and some are found with minor variations throughout Britain and Ireland and North America. The decay of the corpse is a popular 18th-century theme in epitaphs, contrasting the eternal spirit with the ephemeral flesh. There are also many phrases that are placed within the decorative schemes at the tops of headstones that encourage suitably sober thoughts. particularly popular examples include '*Momento Mori*' and '*Tempus Fugit*', but there are also many Biblical phrases and some simple admonitions.

As texts became longer, the emphasis on the body became more explicit. This is normally visible in the opening phrase, and this could be expressed in many ways. These include many variations around the location of the corpse as in 'Here lies the mortal remains', or an explicit link with the memorial, as with 'Underneath this stone'.

2.5.3 Salvation Symbols

The change from an emphasis on mortality to one with a more optimistic view of salvation is one which has been most studied within the New England context, with the winged skull discussed above, and the shift to representations which gave the impression of a face. This then became transformed into the cherub, seen elsewhere such as in Britain and Europe where the winged skull is rare.

In Britain the change from mortality to salvation is less clear than in New England, as in most areas there are few memorials with symbols early enough to show the mortality phase as dominant. The slate headstones of the east Midlands emphasize the cherub, but with significant marginal mortality symbols (Barley, 1948). Slightly later stones in this area include those with Death, but others with the Resurrection (Burgess, 1963).

Resurrection scenes which emphasize judgment on the one hand but also salvation and everlasting life on the other are regularly found but never in any quantity, in Britain and Ireland and North America. Christ in judgment, cherubs blowing trumpets, or figures rising from the ground, may be seen as single elements or all together in a composite scene. Sometimes there is an angel or cherub holding scales of judgment, or a book with a text inscribed upon it indicating judgment (figure 34). The variety of ways in which resurrection and judgment can be portrayed has been discussed by several scholars (Burgess, 1963; Ludwig, 1966), though the full potential of contextualizing these iconographic choices has rarely been attempted. Some show contemporary views of churchyards, with the dead coming out of the ground, and can be informative about graveyard layout and appearance at the time the stones were carved.

Jeanne (1987:69) considers that several of the motifs found on southern folk cemetery memorials reflect a strong fundamentalist Christian view of death and the life to come. Most notable features relate to heaven (finger pointing up to it, representation of the heavenly gates), the Holy Spirit (represented by the dove), the Bible (open book), death and separation (broken link of chain, broken rose), and the cross and the crown. Many 19th-century memorials emphasize a positive outcome, whether through the anchor of hope or the most central Christian symbol, the cross. This reminds the viewer of Christ's sacrifice but also victory over death, and the promise that the faithful will also be saved. Anglicans prohibited the use of the cross on memorials in the 20th century, as its frequency diluted the power of this most important symbol, but its use remained high in Catholic contexts.

2.5.4 Salvation Texts

The Roman Catholic concern for salvation leads to a focus on prayer, which can be readily seen on internal memorials and on external memorials. Phrases such as 'Pray for the Soul of' frequently occur on such memorials of all periods, though most European Catholic memorials only list names and dates. That such memorials served as a vehicle for such prayer can be seen on the 1857 picture of Clonmacnoise by O'Neill. Here, two mourning women were depicted praying in front of a headstone (Manning 1994). Some of the New Orleans tombs, and many in Europe, have seats in front of them which could be used by those praying in front of a tomb (figure 54).

Optimistic texts which imply a pleasant if not ecstatic afterlife are frequent. Phrases such as 'Until the trumpet sounds' are clear, and others such as 'We shall meet again' or even 'Not dead but sleeping' all imply an afterlife where the divisions caused by death shall be ended. Many Biblical quotations are ones taken to give this comforting and optimistic message (Mytum, 1999b); just as the cross symbolically evoked the hope of the resurrection, so the quotation from the New Testament could do the same.

2.5.5 Remembrance Symbols

The final major phase of emphasis on memorials is that of remembrance, where death is no longer greatly emphasized, and can indeed be apparently evaded. Remembrance was of course always an element in the purpose of burial markers, but it became the dominant theme in text and image during the latter part of the 19th century (Burgess, 1963; Tarlow, 1999). This can be seen in colonial cemeteries across the globe and, though manifested in different ways in various countries, was a world-wide phenomenon. In some countries, flowers and appropriate epitaphs were favored (figure 47), in others the use of photographs prevailed. The sentimental motives behind such texts and symbols became remarkably widespread in the late 19th and early 20th century, to be reduced though still present in modified forms during the 20th century and indeed to this day.

From the 19th century the use of kerbs to mark plots became popular, and sometimes they were chosen to be the only element of the monument, and would

Figure 47. Headstone with hand holdiing posy of leaves and flowers, including one with broken stem, Dinas, Pembrokeshire, Wales.

carry the commemorative inscription. The kerbed plot acts in the same way as the earlier monuments as a protecting area, either for individual interments or for family groups. With these later features all reference to the body has gone, but another strategy replaces it, allowing other sentiments to be emphasized on the memorial in text and symbol. The kerbed plot also created a protected area within which more portable tokens of remembrance such as flower vases or immortelles could be placed (see below). Where kerbs supported railings, these items were very secure, though less visible than in a more open plot.

The role of hands has been interpreted by Patterson (1989) in a form of structuralist binary opposition:

ABOVE

PARTED UNITED

BELOW

with different designs and gestures emphasizing one aspect or another. She has illustrated this through a study of some Ontario memorials, but it could have a wider application. It may be difficult to ascertain whether the unstated symbolism was recognized by those carving and using the designs, or whether they had been explicitly designed in one context, and then just copied elsewhere. If this were the case, it would be expected that the copies show inappropriate and perhaps contradictory additions or alterations, or that texts do not support the symbolic messages.

2.5.6 Remembrance Texts

In essence almost all funerary monuments include within them an element which is concerned with remembrance. Some, however, emphasize this over other aspects such as separation, mortality or the afterlife. This may be partly related to fashions of a particular time, but can also indicate individuals' lack of confidence in such theological matters. Remembrance texts concentrate on achievements, status and character of the person when alive, or if they died young, their place related to others of achievement or rank. As has been indicated in other chapters, what is chosen to be remembered about the deceased can be highly informative; here it is sufficient to note that remembrance itself was at times the dominant theme.

3. GRAVE REFURBISHMENT

The grave of the deceased can be visited and looked after whether there is any permanent memorial or not, and some unmarked graves in a burial ground are clearly remembered and tended for decades. The presence of a permanent memorial provides a focus for grieving and remembrance and, in the case of Roman Catholics especially, for prayer. In some periods, regions and cultural groups the burial ground is an important and relevant part of the landscape, much used and cared for; in many others it is marginal and, particularly in recent times, forgotten.

Sociological studies of contemporary grieving and grave visiting in England suggest a relatively short period of intense use, and in many cases a more or less permanent abandonment of grave care after only a few years, though some are cherished for decades (Francis et al., 2000; Hannon, 1990). Indeed, the unofficial elaboration of some grave spaces, with ornaments, figures, fencing and the planting of flowers or shrubs, can cause major problems for those responsible for grounds maintenance. Tensions can therefore arise between grieving families and the authorities, a subject often reported upon in local newspapers. To what extent such conflicts existed in earlier times is less clear. Where these modifications are within a marked plot, such as the plantings of seasonal flowering bedding plants found on many Irish family graves (Mytum, 2003d), there may be less difficulty, though even here they may be against the regulations set out by the burial authorities. In many Catholic countries, items acquired by relatives on pilgrimages may be added to the tomb or placed within a kerbed plot, as aids to prayer in themselves or as evocations to particular saints.

In Britain, but particularly North America, the limited amount of grave visiting is frequently hidden within the larger cemeteries by the purchase of perpetual care agreements which keep graves tidy and even decorated with flowers (Sloane, 1991). Community burial grounds, either maintained by the local authority or church body, may also be well maintained, though without any individual care of particular graves. In Britain many graveyards and parts of cemeteries have been so little maintained that they have become important wildlife sanctuaries, and there is a significant movement to maintain and develop these as a specific form of nature reserve (Barker, 1977; Dennis, 1993; Freed, 2001; Greenoak, 1985). Here the original function is either denied, or the concept of rebirth, peace, and the natural rhythm of life and death are emphasized.

All Saints (or Souls) Day, the 1st of November, has for long been an important festival when in many parts of the world the dead are remembered and graves visited (Florence, 1997), but gradually during the 20th century there has also been an increased level of activity at Christmas and some other festivals in America and elsewhere (Gosnell and Gott, 1989; Hallam and Hockey, 2001:148–151). Many Irish burial grounds still in active use have an annual intensive cleaning, both of the graveyard generally and particular graves and their monuments by family members in particular. Usually the blessing of the graveyard takes place on the Patronal Sunday linked to the dedication of the church. The freshly cleaned and decorated graves show a respect for the past generations and communal solidarity which the many families together represent. There may be an element of competition in cleaning and decorating, but it is the corporate demonstration that is the result. In many Mediterranean countries, tombs and smaller monuments are visited and washed regularly, with the frequent use of artificial flowers not necessarily reflecting intermittent visiting, but rather the harsh climate and the very limited life for cut flowers.

In the American South there is a long tradition of communal burial ground maintenance, where individuals concentrate on their own family plot but also contribute to the general appearance of the cemetery as a whole. This is particularly important where the scraped bare earth appearance requires attention on at least an annual basis to prevent colonization by vegetation (Collins, 1990; Jordan, 1982; Little, 1998). The origins of this phenomenon, and its associated maintenance regimes, is not certain, but it is known from parts of Ireland, Scotland and Northern Europe. In Sweden, for example, the gravely soil within grave plots is neatly raked to provide a tidy and decorative setting for the stone (figure 48). Annual grave visiting also occurs in non-Christian contexts; for example, Chinese families can gather at *Qingming* when they 'sweep the grave' (Chow and Teather, 1998, 289).

Seasonal grave decoration is common in America, where Christmas decorations are frequently placed on the grave, and some are remarkably elaborate. Even in the uncluttered and open lawn cemeteries, wreaths and other decorations are placed on the otherwise visually unobtrusive plaques. This phenomenon is now becoming more common in Britain, though as yet largely without commercial provision of such decorations. Other decoration can be associated with military burials,

Figure 48. Headstones with kerbed plots; the gravel within the plots has been carefully raked to produce a ridged appearance, Uppsala, Sweden.

Figure 49. Left, Headstone illustrated in figure 36 right, prior to cleaning and repainting, Monasterboice, Co. Louth, Ireland. Right, Headstone replaced in 1907 and again in 1970, and still in use, Terrington, North Yorkshire, England.

with the placing of flags on graves in America, and more communal Remembrance Services held at war memorials in many countries, when ceremonies are performed and wreaths laid.

One element of grave refurbishment that needs to be noted is that of renovation of memorials, or their replacement (figure 49, left). This may be a simple repainting of lettering or a more intrusive mechanical cleaning of a stone or re-cutting eroded lettering. In some cases, memorials can be completely replaced, and here the original form may be replicated or a style contemporary with the refurbishment chosen (figure 49, right). Many notable figures whose gravestones have been lost, or who never had a permanent memorial, have had such erected by interest groups in their honor. Such figures are often of cultural or political significance at the time the monument is erected, and so may say more about that period than the one to which the deceased belonged.

In many ways, from individual maintenance and refurbishment, through institutional provision to communal action, graves and monuments are maintained, refurbished and even replaced. This dynamic indicates how burial grounds and the material culture in such locales have a powerful place in people's continuing relationship with the deceased. Combinations of anthropological, sociological and archaeological analysis can explore this relationship and, where appropriate documentary evidence also exists, can consider changes over the long term.

4. CONCLUSIONS

Historic burial sites provide rich evidence concerning attitudes to death, the body, and remembrance. The treatment of the corpse before and after burial can be contrasted with the actions and emotions of remembrance. Some of this information is personal and individual, though displayed in a public setting. It needs to be set against the private forms of commemoration such as photographs, samplers, letters and other momentos in the domestic context that also formed part of personal commemoration. Whilst this may all suggest a highly individualized set of experiences for each death, and this was no doubt felt to be the case to those grieving, the emotions and behaviors were and continue to be culturally situated. Constructed through feelings and actions learnt within the culture, they express normative values which are open to analysis and around which variation can be considered. The burial sites themselves (apart from private family burial grounds) offer insight into more official attitudes, through the organization of burial space and the limitations, if any, placed on modes of commemoration.

Chapter 9

Carrying Out a Study

The recording and analyzing of graveyard memorials can be relatively straightforward and should be non-destructive. It is therefore attractive as a research tool, an educational resource, and a community heritage initiative. This chapter outlines some of the logistical and methodological issues, though it does not provide a detailed step-by-step guide on the recording process for monuments. This needs to be designed for each research or conservation program, and depends on the scale of data collection and the intended methods of analysis (Mytum, 2000). The previous chapters have highlighted topics for research, and the same data is not needed for them all. Here, emphasis is placed on some of the general practical problems of graveyard survey, whatever data is required. Some consideration is also given to some of the problems of excavation, though this is not exhaustive.

1. METHODOLOGICAL ISSUES

Some archaeological studies have seen historic memorials as a form of data on which to experiment and test assumptions regarding material culture that might then be applied to less amenable evidence (Deetz and Dethlefsen, 1965). Most research, however, has focused on death and its associated cultural behavior and beliefs on the one hand, and on social, economic or symbolic deductions which can be drawn from mortuary data on the other. Thematic examination of the material and its interpretation has been discussed in previous chapters; here, emphasis is on the methodologies of placing of data in time and space, and its classification so that, once ordered, trends and associations can be identified.

Gravestones appear to be completely straightforward items of material culture, complete with names, dates, and fixed locations. The burial grounds likewise seem to be unproblematic. Whilst memorials are replete with contextual data compared to most items of material culture studied even by historical archaeologists, the data does not always provide the information it appears to give. The very quality and quantity of information on historic mortuary evidence, combined with additional material and documentary sources, both raises our expectations but also allows for some cross-checking between sources. In this way, confidence can be generated in assumptions regarding the location of memorials and burial grounds in time and space. Moreover, it is possible to consider the relationships between the management of the burial ground, the making and commissioning of a memorial, and the instructions or perceived desires of the deceased. The physical data also has to be placed within some form of classification, which can give primacy to certain variables, of which form and decoration have traditionally been given prominence for both sites and monuments.

1.1 Dating

Mortuary evidence such as burial grounds and memorials provides the archaeologist with many dates. In comparative studies of memorials, dating to a recognised degree of accuracy is essential so that an appropriate form of interpretation can be offered. The majority of memorials include the date of death of more than one individual, and usually have had a complex history as commemorative items. Linked to this, the grave space may have been used a number of times, though not necessarily as indicated on the inscription since memorials may include mention of those buried elsewhere, and other interments in a plot may not have been recorded on the stone marking it. The patterns of grieving, grave furnishing, and visiting will vary according to cultural context, and so the phases of use of a burial ground as a whole will be a complex dynamic. Dating individual monuments, and changes to such memorials, both informs and is informed by the general development of the burial ground as a whole, as it fills up with interments and memorials, and as its vegetative planting matures and is replaced.

1.1.1 Memorials

One of the most important methodological issues regarding monuments is their dating. With their inscribed names and dates, burial markers can give a misleading temporal accuracy. What is inscribed may indeed be true, but inferences we may draw on the data may not be so reliable. The date of death of an individual on a memorial may not have a close association with the erection date of the memorial, or its inscription. For some generalised approaches this may not be a great problem, but in fine-grained studies it can be a major potential weakness. It has therefore been an important area of research where considerable progress has been made in assessing the problem and, in some cases, solving it.

Not all memorials are erected soon after the death of the person commemorated. This is always a potential problem, but it is perhaps most obvious when the headstone was first established in a region. The graves of relatives previously marked by, say, a wooden feature could be commemorated some time after the event in stone, part of a solidifying of tradition and association with place which a number of memorials to members of the same family can help to confirm. Most research into dating has taken place in New England (Slater and Tucker, 1978; Tucker, 1992; Wilson, 1992), but recent work in Britain and Ireland has also considered this problem (Mytum, 2002b).

In New England, the earliest dated stone in one of the major carving traditions, the Essex County style, is to John White who died in 1668, in Haverhill, Massachusetts. Ludwig (1966:358–380) took this to be the date of the stone and thus provided a date for the origin of the tradition. However, further research has highlighted the fact that many memorials were erected a considerable time after the deaths they commemorate, a process which has been termed back–dating. This has been identified as the case at a number of New England burial grounds, and includes a reassessment of the John White stone. Caulfield (1967), followed in greater detail by Slater and Tucker (1978), demonstrated that all the 17th-century decorated stones and some early 18th-century examples belonging to the same tradition as that from Haverhill have all been back-dated. They have thus been able to ascribe headstones with a wide range of dates of the commemorated deceased to a much narrower chronological range of production. Therefore, instead of several craftsmen being needed to cover the long temporal span suggested by Ludwig, the work can all be assigned to the hand of one man, identified as John Hartshorne. Much of this reassessment is based on probate information, as well as typological similarities between memorials. The earliest possible date is now proposed as 1697, though Hartshorne's carving may have begun as late as 1705 (Slater and Tucker, 1978:97). Once back-dating has been used to adjust the headstone erection dates, the data show not only rise and fall in production over time but also John Hartshorne's documented move from Massachusetts to Connecticut.

Other examples of back–dating have been identified in New England. The prolific carver, Thomas Gold of New Haven, Connecticut, changed his style from the 1780s, so that Williams and Williams (1988) infer that a significant number of stones with later features but early dates are back-dated stones because they are in the later style. Sometimes particular documentary sources can be used to show such delays; these may be available in 19th- and 20th-century cemetery records, but occasionally also survive for earlier periods. Using a carver's day book which survives for a five year period Wilson (1992) was able to demonstrate back-dating, and with this detailed source demonstrated that only 19.6% of stones were purchased within a year of the death. Almost two thirds of all stones were purchased within two years of death, which is explained by the delays in resolving estates but also allowing the grave itself to settle. The worrying feature for dating comes from the 14.2% purchased more than five years after the death, and indeed some as much as 15 years later. Wilson (1992:124) considers that all long tails which stretch back in time on battleship graphs are probably the result of this back-dating.

Consideration of dating on British stones has highlighted the many factors that can affect the date of the memorial compared with that of the first death on the stone. Memorials in central Scotland were erected during the 17th and early 18th centuries to mark family plots prior to any burials which were subsequently commemorated on the stone, and the most prominent date on the memorials relates to the original erection, recorded in some cases in documentary sources (Harrison, 1990).

Later monuments can also present difficulties; many were erected only after several family members had died, with their death dates being given in chronological order. Study of the lettering is necessary therefore to determine what has been termed the primary inscription, that part of the text associated with the first group of commemorations, which may be for one person or could be for several. The latest death date within the primary inscription usually provides an approximate date for the stone (Mytum, 2002b). There are also cases of memorials being restored or replaced, or even the inscriptions being removed and the stone re-used (Mytum, 2002a). In Ireland, recent research in County Louth has revealed a number of memorials with dates of erection explicitly stated on the stones (Mytum, 2002b). When these are compared with dates of death for those recorded on the stone, the differences in date can be considered. Here, some memorials were erected years prior to any commemorated deaths, in other cases the delay following death could be considerable. With the 18th-century stones, only half of the sample was erected within a decade of the first death commemorated, though by the late 19th century this was the case with over 80% of the stones in this study.

Many large 19th- and 20th-century monuments have little text inscribed upon them, and many mausolea only have the family name on their exterior (figure 50, left), making dating difficult. Pre-need memorials in recent times also create significant temporal differences between production and death dates (figure 29).

1.1.2 Burial Grounds

Many European burial grounds were established around parish churches during the Middle Ages, so their beginnings have to be established through excavation (Rodwell, 1981). Most urban churches and churchyards went out of use at some point in the 18th or 19th century with the establishment of cemeteries; dates vary from city to city, but it is important to appreciate that burial still continued in family vaults even after these dates (Litten, 1999). Moreover, memorials could be erected within churches and outside them even when burial had moved elsewhere. In English Anglican churchyards, the interment of cremated remains is not considered a burial, and so closed churchyards can be used for this purpose, though permission should be obtained from the Diocese and so should be recorded (Stapleton and Burman, 1988).

Cemeteries, once established, could have remains moved from locations of interment elsewhere. Memorials erected to these individuals can therefore give

Figure 50. Left, Family mauseum made of cast iron with angels on the door, but with only 'Family vault of Camiltenberger' cast at the time of commissioning, La Metarie cemetery, New Orleans, Louisiana. Right, Coffin with the deceased's initials and year of death marked out in upholstery pins, Kellington, North Yorkshire, England.

dates earlier than the foundation of the cemetery (Zelinka, 1991). In other cases, remains were not moved, but family memorials include those already deceased by the time the family plot was purchased, again potentially giving dates for memorials earlier than the opening of the burial ground. The documentary sources associated with local government and commercial cemeteries are often much more detailed than those for other burial grounds, and these can aid many forms of dating and spatial analysis.

Many burial grounds, from urban cemeteries to rural churchyards, have been extended over time (Mytum, 1996). Thus the dating of the burial ground needs to be considered as a phased process, rather than an event. Documentary evidence often survives for 19th- and 20th-century extensions to burial areas, but dating earlier examples can be more problematic. There are usually clear material signs of extensions, in the arrangement of plots and paths, types of vegetation and boundary features, and the dates on the memorials in the various sections of the site.

1.1.3 Burials

The dating of excavated remains is problematic within the historical period. The vast majority are buried with items that provide little indication of a close date. Nail types can be helpful in differentiating between those with older hand-wrought nails and wire nails. It is usually the case, however, that only a few interments may be closely dated within a burial ground, and these may well not be representative of the date range of the site as a whole.

The unusual dated burials can be identified through a range of means. The most secure and effective method is where the date of death of the individual can be read directly off the coffin. Those 18th-century coffins with initials and dates marked out in upholstery pins are particularly useful in providing some fixed points in the early chronology (figure 50, right). Elite coffins in lead often also have incised personal details which help with dating (Litten, 1991; 1999). Trends in coffin type may be identified where the spatial arrangement of graves can be linked to spread in the burial site over time, but these will still be largely relative sequences. In those areas such as Britain and Europe where over-burial was frequent in churchyards, relative sequences can also be established, but at the cost of the loss of much integrity of the earlier burials.

For most interments, the main hope for dating is the survival of a coffin plate. Unfortunately the tin plate that was normally used for these items was thin and corrodes badly in many soils. Not all had text placed on them, but this could be painted or incised. Where these survive then dating is possible. However, the development of an accurate and tight chronological sequence of forms and design motifs may prove difficult. There is some technological and stylistic development which provides a useful generalised framework (Litten, 1985; 1991) but the experience from the Spitalfields study where a large number of dated coffins were recorded during crypt clearance is that many designs continued in use for a long time (Reeve and Adams, 1993). Thus, dated burials of the late 18th and much of the 19th century may help in a site interpretation but may not greatly aid a wider typological study of coffin furniture which could closely date other interments. There are, however, clear differences between early coffin fittings, those of the late 18th and 19th centuries, and the 20th century. Change in coffin and casket production, and the manufacturing methods of fittings, does give the opportunity of dating, linked to producers' and distributors' catalogues (Columbus Coffin Company, 1882; Fellows, nd; Habenstein and Lamars, 1981; Meridan Britannia Company, 1882).

1.2 Location

For many studies the recording and analysis of location is vital. In most cases this is particularly straightforward, but not as often as may be assumed. As with the matter of dating, the evidence presented is not always what it seems. Whilst assumptions about the data may need to be made, where possible this may be usefully cross-checked with other evidence.

1.2.1 Memorials

There has often been the assumption that memorials, once set up, are fixed. Indeed, some early recommendations for archaeological recording suggested the making of very detailed plans so that should excavation take place, it would be possible to link the above ground monument with the correct below ground interment (Jones, 1976). Further survey, documentary research, and some examples of excavation have all indicated that memorials have often shifted small distances from their original siting.

Movement was sometimes part of a re-arrangement by a family on their plot, but it was and is more usually associated with grounds maintenance. Headstones have often been placed in straight lines to aid mowing, and some elements of memorials have often been removed for the same reason—kerbs, foot stones and body stones are the most likely to be cleared away. Sometimes these items can be found discarded in a little-visited area, clearly out of position, but some may be re-set elsewhere or buried. Table and chest tombs may have been reduced to look like ledger stones when the lower parts of the memorial had become unsafe, and in the process may be moved to a different location. Tops of tombs, ledgers and headstones may be used as paving, though some ledgers do mark original brick-lined graves placed along paths so that they would be noticed in the same way as internal burial along church aisles could have been. In the most drastic clearances, memorials may be buried beneath the grass, or placed around the boundaries of the churchyard or against the church wall, though in some areas such as parts of Britain, a few memorials were originally placed in such external locations.

Reliability of location will matter to varying degrees depending on the type of study. In some cases the exact position is important where intra-site patterning and associations are vital, as with Brown's study of the Coolidge family monuments in Watertown, Massachusetts burial ground (Brown, 1993). For many analytical purposes, however, memorials can be assumed to be in their original positions, though other memorials may have been completely removed and the sample now available is no longer complete. Most memorials have remained within the burial ground, and regional studies can be generally confident regarding distributions. Occasionally stones have been moved from elsewhere, as with the transfer of remains and memorials when churchyards have closed, as with Devonshire Street Cemetery, Sydney, and their move to Bunnerong Cemetery (Zelinka, 1991). Usually such moved stones are either documented, or placed in a discrete area which suggests an abnormal history. The example of the Edmond Perkins stone, probably erected in Boston but ending up set in a chimney in Cape Cod, is an example of a movement away from a burial context (Chase and Gabel, 1997:523). Other examples involving the reuse of cast iron monuments in England are noted by Willatts (1987:110). Most reuse of memorials in buildings seems to have taken place within burial grounds, in the construction of boundary walls and paths. The rarity with which memorials are found in other structures suggests that outward movement was unusual, though theft of statuary and elements such as plain slabs and kerbing is well known from modern cemeteries, as such materials can be easily reworked. Moreover, the removal of subsidiary elements of graves such as kerbs, body stones and railings in maintenance of graveyards may lead to their removal and reuse.

Loss rates of stones can be difficult to calculate, but Williams (1994:79) considers that only about 20% of Solomon Brewer's output from the late 18th and early 19th century now exists. In many burial grounds survival would seem to be much higher than this, particularly from the later 19th century, but attrition through vandalism, over-zealous management, and even theft, can have its effect. A combination of oral history, any local documentation (including photographs), and close examination of the condition and placement of the memorials in the ground should provide a basis for evaluation. Loss rates can be considered when the numbers of broken, eroded and moved stones can be calculated, and documented and oral sources concerning site management may reveal information about phases of clearance, reorganisation and repair. Fallen and buried stones may be located by various forms of survey (see 3.1 and 3.2 below).

1.2.2 Burial grounds

The study of burial ground location requires contemporary data from archaeological or documentary sources. Successful studies have considered the location of family burial grounds in relation to the farm houses where those individuals had lived (Gibb, 1996), and how communal burial grounds shifted over time (Brooke, 1988). Studies of urban cemetery locations have been less substantial, with an emphasis so far on land use issues (Hardwick et al., 1971; Pattison, 1955), though in colonial situations some political and cultural factors have been considered (Christopher, 1995; Yeoh, 1991; Yeoh and Hui, 1995). Rural graveyard locations may be linked to religious buildings such as churches, but may be separated from them. A combination of chronological, denominational and ethnic factors may have an impact, as well as topography and land use when the sites were established. When sites became full, and an extension or additional new site was required, these factors again came into play, and are worthy of consideration. In urban contexts, land use was a major factor in location, with a frequent pattern of cemetery establishment on the edges of cities, with these sites subsequently surrounded by suburbs. The chronology and location of burial within a city as a whole can throw light on land use history more generally, as well as strategies for body disposal and commemoration. The urban distributions of monumental masons and funeral directors also show shifts over time, but have been little considered.

1.3 Classification of Memorials

One of the first uses of gravestone data was in the examination of the principles of seriation by examining changes in monument decoration over time and space (Dethlefsen and Deetz, 1966). Work has continued on this aspect, though there is still much to be understood. The data can be interpreted at a number of levels, from that of the individual mason and a workshop's products, to widespread cultural change within a region or even on a global scale.

1.3.1 Shape

Classification by shape has often been undermined by the large degree of variability within and between forms. Many of the problems inherent in typological classification of cultural material within archaeology can be amply demonstrated in a graveyard. Whilst this may be a valuable pedagogic tool, this is little comfort when a research programme is required. There has also been some difficulty in separating out form from decoration on more complex memorials. Only headstones have been subject to widespread analysis for the 17th and 18th centuries (Burgess, 1963; Slater, 1996a; Tucker, 1992), with only isolated and regional classifications of ledgers (Chater, 1976; 1977) and pedestal tombs (Elliott, 1978). Art-historical categorisation provides some framework, for example using revival styles (McDowell and Meyer, 1994), but has not been undertaken on a basis which allows for any quantitative evaluation of trends.

For 19th- and 20th-century memorials, simple classifications have been produced to provide sufficiently large numbers for seriation, but this can lose much of the texture which has been recognised in the earlier stones. There has been an assumption that such material is somehow more uniform, yet in fact this is certainly not the case. Cannon (1989) used broad categories for his survey in Cambridgeshire, England: simple, complex and top-piece headstones, three-dimensional, and coped monuments, together with chest and pedestal tombs, but this hides a plethora of variability within all these. Norris (1988) also utilised a simple classification for Ontario memorials. Although some elements of height and decoration were recorded, associations between these variables were not published. Dethlefsen (1981) has also provided simple classifications for monument types from Florida cemeteries, though more categories are given.

The level of detail for any classificatory system depends upon its research purpose, but many have been at a simple level to date. A hierarchical recording structure has been designed which allows the various elements of the memorial to be recorded separately, and thus many combinations of elements of both form and decoration can be identified and subsequently analysed (Mytum, 2000). Many aspects of memorials are clearly regional, even at a late period, and a system must be designed with flexibility. A structure which allows large scale grouping together for broad trends, but also finer-grained analysis to allow the recognition of regional or local traditions, is valuable. It also allows an assessment of the degree of variability over time which the broad categorisations often inevitably hide.

1.3.2 Decoration

The ways in which monuments are decorated are so numerous as to defy a single overarching classification. Both form and style play a major role, and the relative importance of different motifs and their relative location on the stone can all be meaningful. Many local, regional and wider temporal cultural norms can be noted on visiting burial grounds, and the grammars of layout and the reasons for choices of motifs within these can be studied if desired. Text is also in part decorative, in its font, size and location on the memorial. To record many of these aspects in coded form becomes unwieldy, though some categorisation is necessary (Mytum, 2000). Much reliance in the end may be placed on comparison of photographs during post-fieldwork analysis. Intuitive study of the monuments in situ, or of photographs and rubbings, can allow the human eye to identify key features of style in motifs or lettering that can be subsequently explicitly defined and consistently recorded. Fortunately, it is easy to revisit memorials to enhance data at a later stage when these detailed characetristics have been identified.

1.3.3 Other Attributes

Many other attributes of memorials, such as materials, may have more than one state, but these are rarely as complex as those associated with decoration. The extent to which these may be recorded will vary according to the project design. Aspects which have received very little attention to date are the constructional methods of memorials, the ways in which they have been repaired, and the role of ephemeral and subsidiary elements such as body and foot stones, flower vases, fencing or railings, and other decorative features. They have been noted in many studies, but not systematically analysed and their role in the grieving process has been rarely considered.

1.4 Demography

There is now a well developed set of methodologies related to the study of human remains (Brothwell, 1981; Mays, 1998; Ubelaker, 1989; 1995), and these have been frequently applied to those of the historic period, particularly in

North America (Grauer, 1995; Saunders and Herring, 1995; Poirier and Bellantoni, 1997a). Whilst general demographic trends may be recovered through documentary sources, the demographic patterns from excavated human remains can be combined with any palaeopathology and the material culture of burial to create a richer and more contextualised study (Cox and Mays, 2000). Many palaeopathological and demographic studies can be found in journals such as the *American Journal of Physical Anthropology* and the *Journal of Human Evolution.*

The study of the remains from the Spitalfields crypts has been particularly important because of the named 18th- and 19th-century individuals which could be linked to a rich historical record and well preserved coffin furniture and contents (Molleson and Cox, 1993). Early North American assemblages may be linked to farmsteads (Gibb, 1996), and later populations have been associated with rural contexts where redevelopment or disaster has caused disturbance (Bellantoni et al., 1997; Slkedzic and Hunt, 1997), though few have been published and most remain in the grey literature. Other specific populations have included those from almshouses (Phillips, 1997). African American samples have been derived from plantations (Handler and Lange, 1978; Khudabux, 1999) and urban settings (Crist et al., 1997). Expected patterns of behaviour with regard to burial and commemorative regulations have been investigated through excavated burial grounds and monuments belonging to the Society of Friends, the Quakers (Stock, 1998a). This has led to reassessment of the ways in which Quakers conformed to their rules, and the variability between Quaker communities (Bashford and Pollard, 1998; Stock, 1998b).

Some attempts have been made to use graveyard monument data for demographic purposes (Dethlefsen, 1969; Stewart, 1978; Osborne, 1974). However, this is an unreliable method (Husbands, 1980), as there are many factors which have affected the percentage of the population with marked graves, and the social background of those so commemorated is not a simple reflection of the whole population either. Burial register data, where available, can be used to compare this pattern of monumentality, and changes over time and space can be identified (Mytum, 2002e). This should be carried out in many more contexts to identify patterns in commemorative practice, and more detailed contextualised study incorporating census and other data can give ideological and social data for the deceased.

2. PROJECT FIELDWORK PLANNING

There are many practical issues to consider when planning and carrying out a field project at historic burial grounds. Many are those found at other archaeological sites, but there are unique features which need special attention, and these are the ones focused upon here.

2.1 Types of Study

Studies of burial grounds fall into three main categories, each with very different project implications. The first relates to a survey of a burial ground, perhaps as part of a management plan (Lavelle, 1992:19–20) or in advance of potential development. In this case the definition of the extent and nature of the site is of critical importance, together with some assessment of its research potential. How this definition can be achieved is discussed below, and previous chapters indicate areas of research potential for such sites. Where the need for conservation of memorials or other cemetery features is necessary, many of the methods relate to historic building conservation, though some aspects require more specialist attention (Hosley and Holcombe, 1994; Maxwell et al., 2001; Strangstad, 1988). Again, detailed records should be kept of all conservation activities.

The second category of study relates to research of the upstanding remains in the burial ground, with more detailed and focused recording and analysis of selected evidence from one or more sites. This may develop from the first type of study, or be undertaken by researchers at any educational level for their own purposes. It may have any combination of archaeological, anthropological, sociological, art historical, cultural geographical, geological, genealogical, or historical emphases, but many of the practical issues will be similar and are discussed here. Some of the aspects of recording will not be appropriate to all projects, but these can be selected by the researcher as required. As this form of research is non-destructive, selectivity in the focus of fieldwork is fully acceptable.

The third category of study involves the excavation of graves, either for research purposes or, in most cases today, in some form of mitigation strategy. Most excavation procedures on burial grounds are similar to those on other sites and relate to national and regional styles of excavation, particular research questions, levels of resourcing and time available. There are, however, specific issues regarding excavation of human remains which need to be explored. These include legal, safety and ethical issues as well as methodological and research priorities.

2.2 Preparatory Work

The location of many historic burial grounds may be discovered through use of maps, aerial photographs and other documentary sources. Government or religious bodies may have some or all such sites listed, though the size and density of such sites varies considerably from region to region (Zelinsky, 1976:177; 1994). The range of documentary and other sources will vary greatly from one country or region to another (Bell, 1997; Mytum, 2000). Most cemeteries will be immediately obvious on the ground because of the presence of grave markers, or clear depressions or mounds about the length and width of an interment. Whilst most communal burial grounds will have clear surviving boundaries, the extent of long-abandoned cemeteries and small family graveyards may be less clear. Vegetation may provide a clue, such as the Bois-de-Arc trees often used to define burial grounds in parts of Texas (Owsley et al., 1997), or the cedars and black locust trees in Illinois (Price, 1966). In overgrown areas finding sites at all can be difficult and, even in apparently well maintained graveyards, some sections may be overgrown either through neglect or as part of a deliberate wildlife conservation strategy.

2.2.1 Preliminary Assessment

The choice of burial ground to record may be based on management or pure research priorities, but there are several preliminary assessments that need to be undertaken to ensure a successful project. The number and condition of burial plots and memorials needs to be assessed, and all areas of the burial ground, even overgrown ones, need to be checked. The range of memorial types and date range should also be evaluated, as this will affect the research questions, form of recording, sampling strategy, and logistics of the project. It may be relevant to note unmarked graves, and also other features, and whether memorials have been moved or elements such as kerbs or footstones tidied away. Fallen or buried memorials may also be present, often indicated by vegetation changes.

Documentary research is desirable to obtain some sense of site chronology independent of the memorials, as they may only cover a short part of the period of use; these may include burial registers for the site, and for some graveyards these extend back into the 17th century. Moreover, information regarding the status of the burial ground and its occupants will inform the research and recording priorities. Old maps may indicate changes in the shape and size of the burial ground, and photographs can show not only the presence of memorials but also planting schemes, management regimes, and events such as funerals and the tending of graves.

In some regions there is already a long-standing tradition of graveyard recording, though this may have only involved collection of genealogical data. Some state heritage records may include graveyard data, and in the United States of America they may have been placed on a state's Master State File with supporting documentation (Thompson, 1989). Clearly such evidence can help refine research aims, and may allow integration of already collected data with that from new fieldwork. It is often more difficult than might be expected to integrate existing sources, as not all memorials may now be identifiable, and others may have been previously missed or incorrectly transcribed. Local informants may also be an important source, and may give details which are lacking from the official records.

2.2.2 Permission

Whilst family burial grounds may be privately owned, most cemeteries are owned by a corporation or local government. Permission for a preliminary visit is rarely required for public or corporation burial grounds, as they are usually viewed as spaces open to the public, though this is not the case for family burial grounds on private land. Moreover, detailed study on any site requires notification of the relevant authorities and the gaining of permission. Graveyards may be owned by local government, a private company or trust, or a religious denomination. Most burial grounds in use have a notice board near the entrance giving details of ownership and management. It is desirable to obtain written permission, so that this can be shown to anyone in authority who visits the site. In requesting permission it is important to outline the type of work undertaken, and its final purpose. The degree to which vegetation may be cleared, or surface deposits moved so that lower parts of memorials can be read, needs to be made clear. There may be constraints on taking rubbings of monuments. It is useful to agree what copies of the primary record, and any subsequent reports, are required by the responsible authorities, and to whom they should be sent. Legal requirements may require permission from other agencies whenever excavation is undertaken, and this is discussed in the appropriate section below.

Timing may also be an important matter to discuss with the managing authorities. For example, recording around a parish church may be considered inappropriate around the times when services are being held. In burial grounds still in use, no recording would normally be expected in the vicinity of interments whilst they were being carried out. These are often by necessity arranged at short notice, so this needs to be confirmed by an office visit or telephone call immediately before work begins and, on a longer term project, at regular intervals. This also gives the opportunity to update the site managers and keep them involved, and deal with minor queries regarding permissions and problem solving which often emerge as the survey is carried out.

2.3 Health and Safety

Conducting survey in burial grounds can carry the same risks as any fieldwork, and these vary from one environment to another. Health and safety may be particularly important if excavation is taking place, or survey is undertaken within crypts or burial vaults (Kneller, 1998; Young, 1998; Thompson, 1998; Ubelaker, 1995). Particular care needs to be taken by those conducting survey alone, as sites can be isolated and rarely visited. Uneven ground due to grave subsidence can be an unexpected hazard, as can eroded or vandalised memorials which can be liable to collapse (figure 51). Particular care should be taken in overgrown areas. Personal safety can be a factor in some cemeteries as assaults are not unknown.

Figure 51. Left, Family plot with railings in which a burial shaft has collapsed, creating a void, Church Street graveyard, Mobile, Alabama. Right, Headstone made of many pieces of slate in advanced state of decay, Newport, Pembrokeshire, Wales.

2.4 Etiquette in the Burial Ground

Burial grounds are landscapes loaded with emotion, and the conduct of any project should be appropriate. In zones where graves are still visited, particular sensitivity is required. Survey equipment should not be set up directly on graves, and grid pegs should not be large or driven far into the ground. It may be necessary to minimise walking on grave plots, and recording equipment such as clip boards and tapes should not be left lying around. In many burial grounds it could be deemed inappropriate to have drinks or meal breaks, and even smoking may be thought inappropriate. These issues need to be considered in the context of each project. Where large teams are carrying out recording, such as in a school or college project, then noise levels need to be kept low, particularly in a cemetery in use. It may be better to have shorter working periods to maintain concentration than longer stretches when some recorders may become distracted and indulge in inappropriate behaviour.

It is usually helpful to engage those visiting graves in conversation if they are curious about the recording activity. In almost all cases, the bereaved are delighted that interest is being shown in the burial ground, that records are being made of the monuments, and that this place so special to them is recognised by others. Often, important information about the history and management of the burial ground can be obtained from visitors. Those contemplating studies incorporating more systematic data from visitor informants need to give particular thought to how such data could be collected sensitively. The literature in anthropology and sociology is rich in methodological guides, and the sorts of results obtained from such studies are also numerous (Bradbury, 1999; Howarth and Jupp, 1997 and the journals *Mortality* and *Omega*).

3. SAMPLING

The size and nature of the data set to be collected and analysed is a crucial matter with regard to graveyard recording. However, unlike most other archaeological fieldwork, it is possible to assess the nature of the resource before undertaking concentrated fieldwork. Sampling may be based on all memorials from a single site or group of sites, but analysis may also be undertaken from a smaller sample. The population to be sampled may be the whole assemblage in a site or region, but may be more focused, for example typologically or chronologically. Moreover, if changes in style over time are to be analysed, it is relatively easy to continue to acquire relevant data from certain times to ensure that the whole period under consideration is provided with sufficient numbers of examples, ignoring redundant data from those time periods for which sufficient has been collected, and concentrating on augmenting the sample size for those still under-represented. Likewise a more extensive recording programme for rarer monument types such as tombs can be carried out to allow for the analysis of changes in their form over time, when the collection of headstone data may be more rapidly acquired. Clearly if the proportion of headstone to chest tombs was the research focus, the the relative rarity of one form over another would be a vital aspect of the research, and the sample would be of all monuments, though still perhaps modified to give an even temporal spread.

3.1 Spatial Sampling

Many burial grounds are spatially segregated, based on denomination, social factors, or chronology as the burial area was extended. These factors need to be evaluated from documentary and reconnaissance survey before an appropriate sampling strategy is devised. Samples may be taken from only one of these sub-populations, from many in appropriate proportions to represent the whole, or as separate populations to be compared and contrasted.

Methods of obtaining samples vary. Effective methods include using selected zones already defined in the cemetery, such as blocks defined by paths. Alternatively, recording every fifth stone for a 20% sample, by moving along the rows, can allow a spatially more distributed coverage. Random collection can be time-consuming unless a numbered plan already exists; given the amount of prior knowledge about the cemetery gained from preliminary research and reconnaissance, a more efficient stratified sampling strategy would normally be preferred.

3.2 Sample Size versus Sample Detail

There are important decisions to be made regarding the number of categories of data to collect for each memorial, and the detail to which these are recorded. These factors have to be traded against sample size given a finite survey resource, and are determined by the purpose and context of the survey. Some of the issues regarding research questions and design of recording forms are discussed in section 5.1 below.

Where survey is being carried out for general use by a variety of potential researchers, or within CRM contracts, a wide range of information at a low level of detail may be required from many or all memorials. In contrast, research concentrating on particular questions may be based on a more limited but highly selected sample and with only relevant variables collected. Where genealogical or local history groups have already collected data on the texts, this may be incorporated within a material culture data collection sampling procedure, and the existing transcriptions merely checked against the memorials. Such sources may also allow judgement on sampling policies within and between burial grounds, within a programme of field visits to assess survival, accessibility and the material culture variables appropriate for recording.

3.3 Student Projects

Graveyard surveys form excellent student projects, as the data is easily available, is present in all regions, and can be collected relatively quickly even by a single individual. It is possible to ask either methodological questions relevant to a discipline (such as typology and distribution), or particular cultural or historical questions. It is important, however, that students fully consider their sampling strategies, otherwise collection leads to large data sets that cannot be easily analysed, or the data has come from an undefined population. Students working in teams within a strategy devised by teaching staff can be very effective, but when working alone there is the risk that much effort and enthusiasm will be wasted. When undertaking projects independently, on their own or in groups, students need to be aware of health and safety issues, keep responsible people informed of their movements, and where possible have a cell phone for making contact in case of emergency. In some urban cemeteries, problems of personal safety are particularly acute, particularly in the older, less-visited areas.

Students are particularly prone to under-research existing resources, such as inscription transcripts and documentary sources about a burial ground, before planning and carrying out their fieldwork. Appropriate advice is therefore necessary to prepare students before setting out on their research. The categories of data, and their level of detail, are also aspects on which students often need advice. It is easy to collect too many data categories not subsequently used in analysis, or to have too much field data that is then unwieldy to sort and analyse, whether manually or by computer. However, the results from student work can be extremely impressive, either in reinforcing existing knowledge of commemorative practice, or in exploring specific developments and their causes in a particular locality.

3.4 Excavation

Only in exceptional circumstances will there be any choice regarding sample size in historic cemetery excavations, as few are purely research projects. Normally excavations are related to the removal of burials for reinterment, dictated by the scale of threatened non-archaeological disturbance. Nevertheless, the level of analysis of both anatomical and cultural material possible from the evidence collected will be dictated by sample size. These methodological issues are not restricted to historical sites, though the possibility of linking documented familial data to personally identified interments gives particular opportunities for some research only otherwise possible on contemporary populations.

4. SURVEY

Many studies of burial grounds will require a map to show spatial patterns and the location of particular monuments or burials. Cemeteries in use will usually already have accurate maps used in their management, particularly where plots are to be sold. Older graveyards may only have some schematic representation, which may be incomplete, or no map at all. Most abandoned burial grounds will have no surviving map, though commercial and local government cemeteries would once have had such records, and the appropriate archive repositories should be consulted. Even where there are maps, however, these may only show plots or mark all burial locations regardless of whether there is any monument or not. It may be possible to annotate a copy of an existing map, but in many cases the researcher will need to produce their own. Depending on the research purpose, this can vary in quality from a sketch plan to a highly detailed study. Suggestions given below indicate the types of decisions which need to be made before or during the survey.

4.1 Surface Mapping

For some purposes a sketch map is sufficient, to allow subsequent location of memorials during fieldwork. Even on such a map, features such as buildings, boundary features, paths and identifiable trees and bushes should be marked and annotated. It can be useful to indicate the types of monument by symbols so that it is easy to locate them. Annotating some memorials with the names of those commemorated also aids recognition, and is helpful when the sketch map is being checked for duplicate or missing stones.

4.1.1 What to Plan

It is vital that there is a clear policy on what is to be recorded, and what drawing conventions are to be used. The policy should be written down, kept with each copy of the map, and incorporated into any final report. Monuments may be marked simply by a dot, an outline, or more detailed shape. Usually kerbs and railing should be marked, as should unmarked burials where they can be identified. Some burial grounds may contain reused objects as burial markers, so care has to be taken to distinguish between these and refuse.

It is usually necessary to number the memorials, if these are to be individually recorded. In some cases, unmarked burials may also be numbered; they can be included in one running list, or given a separate set of numbers. If only a sample is to be researched, this may be derived from the fully planned and numbered total, or only those in the sample may be numbered. Numbering should start at an obvious point, such as at the graveyard entrance or the corner of each zone of burials in a cemetery. Where there are rows these can be followed, but often markers are more erratically arranged. Numbering can take place before a detailed plan is made, particularly if detailed recording of memorials is to take place before or during this mapping, but then a sketch plan must be made, otherwise confusion will result. White plastic tags marked with numbers may be placed by memorials to aid identification, but these are only useful in sheltered, rarely visited locations. Marking the backs of well preserved memorials with chalk may be viable in dry conditions in abandoned burial grounds, particularly those that are overgrown, but is otherwise inappropriate.

Many large cemeteries have elaborate numbering systems for their plots, and these may be recognised both on maps and on the ground where the memorials themselves or special markers set in the ground may be inscribed with a code. This may be quite complex, involving letters and numbers indicating zone, row and plot. It is usually unwise to use the cemetery numbering system as the basis of a record, as many plots and memorials will not be marked, and working these out can lead to errors and confusion. It can also be difficult to sort cemetery numbering systems by computer, but it is helpful to record them on one of the data fields as in later integration of documentary sources and the material record they may be invaluable.

All visible boundaries should be plotted, and where construction varies this should also be indicated. Many features such as paths, roads, landscape features such as rock outcrops, ponds and vegetation are useful to indicate the character of the burial ground, aid orientation on the ground, and can be important for interpretation. Annotation on the map or records of another form may be required indicating whether topographical features are natural, modified or man-made. It is often important to note the species, height and approximate age of trees, and whether they were part of a deliberate planting or are natural regeneration. Other vegetation may be significant, being either deliberate plantings on graves (which may be otherwise unmarked) or reflecting natural colonisation of disturbed ground.

4.1.2 Carrying out the Mapping

For archaeological publication and even for archive deposition, a map more accurate than a preliminary sketch is usually necessary, as sketch maps are often difficult for others to interpret. The final survey plan should normally be at a scale of 1:100 to allow enough detail, but on a large cemetery scales up to 1:500 may be necessary. It is better if at all possible to carry out surveys of large areas in sections at a scale of 1:50 if there are complex monuments. In this way, sufficient detail can be retained, and smaller scale overall plans produced from these as required. A list of suggested equipment is given in Table 3.

Smaller graveyards or sections of large cemeteries can be surveyed by a team of 2 or 3 people using tapes. A base line should be set out across the survey area, from which grid squares marked by pegs can be set out. From the sides of the grid, offset measurements can be taken to particular graves and monuments. In well ordered cemeteries, it may be easier to plot the ends of rows and then measure along them, even though they are not on the grid. Sites on a steep slope or with much vegetation cover can be very hard to survey, and a lower standard of accuracy may have to be accepted in these situations.

Contours can be added to a survey made with tapes and grids by using a level (Mytum, 2000). There will normally be large numbers of identifiable points where readings can be taken, next to memorials, on paths, and by trees. It may be necessary to take some readings in areas where there are few upstanding features. Intensive surface survey is now easily possible with total station theodolites and the appropriate software. Slight depressions (caused by decay of coffin voids and bodies) and low mounds (produced by the unconsolidated grave backfill or still surviving coffin) can be detected through such survey if the density of readings is great enough. Overgrown pathways, landscaping features, and subdivisions within the burial ground can also be recognised. For such features to be visible, a plot scale of 1:100 or even 1:50 may be necessary. Computer-generated oblique

Table 3. Checklist of equipment for fieldwork

Basic Memorial Recording—one set for each Team Member
Blank Record forms
Code sheets
Clip board
Medium pencil (HB)
Pencil sharpener
Soft eraser
Hand tape (3m with grip)
Flashlight
Soft brush
Plastic garden tag or lollipop stick
Plastic ruler
Basic Memorial Recording—for each Team Member or to share
Numbered sketch plan
Mirror or board covered with tin foil
Compass
Tools for vegetation clearance
Spray bottle
Sponge
Trowel
First Aid kit
Planning Team
TST and prism *or* Survey tapes (30m or 50m preferred)
Compass
Surveyors' arrows or equivalent
Grid pegs (e.g. heavy duty plastic tent pegs)
Board covered with graph paper (mm and cm squares)
Plastic drafting film
Masking tape
Hard pencil (6H)
Pencil sharpener
Soft eraser
Plastic ruler
And, if memorial recording has already begun:
Numbered sketch plan
Record forms to check identification
PhotographyTeam
35 mm camera (SLR) & print film *and/or*
Digital camera & Laptop for downloading images
Scale (30cm or 50cm and 1m)
Pegboard and letters, numbers *or* Blackboard, chalk, cloth
Flash on extension
Mirror or board covered with tin foil
Tools for vegetation clearance
Soft brush
Graveyard plan
Record forms to check identification
Rubbings Team
Wax crayons or equivalent
Rubbing paper (e.g. wallpaper lining paper)
Elastic ribbon to hold paper in place (2cm width)
Drawing board on which to lay out rubbings to check in front of stone
Masking tape to fix rubbing to board
Soft pencil to clarify rubbings during checking (B)
Graveyard plan (sketch or measured)
Record forms to check identification

views of intensively surveyed burial areas can reveal features difficult to see on simple contour plots, particularly where the ground is naturally sloping or where vegetation hides the variation in ground surface.

The use of GIS can allow the memorials, topography and other features such as paths, vegetation, and boundary walls to be placed on different layers. If each memorial has a layer of its own, and the database of memorial information is linked to it, then more complex spatial aspects of temporal and social trends can easily be investigated (Mytum 1996).

4.2 Geophysical Survey

Geophysical prospection can be effective (King et al., 1993; Owsley et al., 1997), and there has been considerable experimentation with soil resistance (Ellwood, 1990), magnetometry (Ellwood et al., 1994), and some use of ground penetrating radar. Much depends on the intensity of burial, the nature of the subsoil, and contents of grave fills. Where there has been much intercutting or build-up of deposits over burials, it may be very difficult to identify individual graves. Areas of ground disturbance may be located in these circumstances, but only excavation can at present define the graves. Brick or stone lined vaults and burial shafts may be relatively easily located through geophysical prospection, as may those graves with many metal fittings or cast iron coffins. Paths and other landscape features may also be identified through survey. Burial ground boundaries marked by fencing using metal fittings can be effectively located with magnetometry (Owsley et al., 1997:207–208). Surface and subsurface survey methods may positively identify graves, but they rarely if ever can demonstrate conclusively that there are no burials in an area.

Geophysical survey areas should be accurately plotted and related to easily identifiable surface features. A full methodological report should be provided, with equipment specifications and the conditions under which the survey was undertaken. Factors such as the transect interval and number of readings per metre are critical for the reinterpretation of data.

It is possible to use probes and (with permission) augers to locate otherwise invisible graves or buried memorials. These low-technology methods, and indeed the thumping of the ground with a wooden or metal pole, can be effective in locating areas of less compacted soil or even voids.

4.3 Photography

Whilst much photography is concerned with recording individual monuments (see below) it is highly desirable that surface or geophysical survey is augmented by photographs. These can be an important part of the archive, to be kept with the survey methodology report. General photographs show topography, vegetation, and management practice at the time of the survey; moreover, a selection of memorials can be photographed to provide a key to the monument types used on the plan. The location from which general photographs were taken, and the direction faced, can be marked on a copy of the plan. The list of equipment needed is given in Table 3.

Aerial photographs of a burial ground can also be a valuable record, provided that they are taken from a low enough altitude that some detail can be seen on the image. Photographs may need to be taken at different times of the year to take advantage of lighting conditions and extent of vegetation cover.

5. RESEARCHING MEMORIALS

The advantage of researching upstanding remains is that a refined and effective research design can be prepared at an early stage, and its implementation carefully considered. It is also possible to return to the memorials to collect additional data as required. Many community projects can be built around memorial recording, and can incorporate oral history and opinions from a wide range of social classes and age ranges. The type of information recorded, the format of the record, and the nature of the analysis will vary depending on the interests of the participants and the intended output of the results. Studies can be built around one particular graveyard or group of graveyards, or a monument form or a carver's work over a region.

5.1 Written Recording

It is highly likely that some type of form will be required to collect standardised data. Most of the forms which have been published assume a generalised recording purpose rather than a tightly defined research aim, and each memorial has a form (Jones, 1976; Mytum, 2000). Some papers have published the form used to collect the data (Norris, 1988; Shoesmith, 1980). On a research project often only a limited range of data needs to be collected, and here each record sheet could be used to store data for a number of memorials. Forms for school children can be adapted for the age range and abilities of the students (Mytum, 2000:160), with drawings instead of written codes, and large enough boxes for entry of free text such as names. Some examples of forms are given in Appendix 1.

5.1.1 Logistics

Projects undertaken by a single person need to be carefully defined as it is easy to collect too much information from each memorial and not obtain a large enough sample for analysis. Collecting even a limited range of information about each stone may only allow 10–20 well preserved memorials to be recorded a day, though much varies according to field conditions and the legibility of the inscriptions. With group work, it is essential that allocation of memorials for recording is coordinated so that some are not omitted and others recorded twice, and if a sampling procedure is in place this needs to be regularly monitored. Recorders can work alone or in pairs. Where there is vegetation clearance, removal of deposits around the foot of stones, the taking of photographs and rubbings, as well as written documentation, these activities may be undertaken by different team members, by everyone as required, or in stages. The basic equipment needed for many surveys is listed in Table 3.

5.1.2 Reading the Inscription

Whilst many memorials are easy to read, others can have eroded badly, or be covered with lichen or moss which impedes legibility. Moreover, older memorials may be carved in unusual letter forms, have unfamiliar words or phrases, and use enigmatic abbreviations. Various methods have been developed to improve the recorder's ability to read a stone, though undoubtedly the most important is that of experience. Over time, researchers in a particular area become familiar with the local styles and can recognise whole sections of text from even fragmentary inscriptions. Many introductory phrases are in elaborate script, which often confuses the beginner. It is important to realise that the names of carvers and masons, in full or abbreviated form, may be found on a minority of memorials. The most common place is near the bottom front face, but they can appear in various places on the stone, and even within the design.

Texts and indeed incised decoration can often be enhanced by careful use of a soft brush, though this must not damage the surface of the stone. Wetting the surface can highlight fine lines, either immediately following application or as the surface dries out. This can be achieved with a fine hand spray (such as those used for houseplants), or if more water is required, this can be applied with a sponge. It is undesirable to wash or brush a memorial any more than is necessary to read the stone; too much cleaning can have a deleterious effect on the stone and can greatly increase the rate of weathering. Likewise, clinging vegetation should only have leaves removed, and lichen should be left untouched. Moss may be removed with care on stones with a solid surface.

Good lighting is the most important factor in reading a stone, and oblique light improves legibility immensely. The use of a flashlight, or deflected light from a mirror or a board covered with aluminium foil, can be very effective, particularly if moved back and forth to highlight the letter forms at varying angles. Such methods can also be valuable in identifying decorative features.

Where inscriptions are being recorded, it is difficult to transcribe the style and size of text, though upper and lower case, and superscripts, can be written down as on the memorial, and italics can be indicated through underlining. Rubbings or photographs can be used to record accurately and consistently text styles and sizes if these are considered important. The ends of lines of text can be indicated by the use of /, but should normally be set out on the form as on the stone, which can also then indicate whether text is centered or not.

All recording projects in burial grounds require careful checking. Not only are inscriptions often difficult to read, but the language, spelling and terminology can be unfamiliar and easily misinterpreted. Some categories of data are difficult to code consistently, and errors in form filling are surprisingly common. It is easy for recorders to omit data or to write it in the wrong place on the form, and some writing can be less legible than others. Data input to computer can also introduce further errors. A system of field and base checks should be part of the process for team projects, and single recorders should carry out some checking of their own data. Sufficient time should be set aside for field checking, as this can be a relatively time-consuming process but essential for the production of accurate and consistent data.

5.1.3 Coded Information

Much information can be coded on the field record form. As with all forms of classification, this can be based on a series of broad headings (Cannon, 1989; Dethlefsen, 1981), or can have a hierarchical classification system which allows for the breaking down of the data into more detailed categories, or restructured according to different criteria (Mytum, 2000). The research questions dictate the degree of detail, and this applies to categories of data as varied as geology, monument form, and decorative motifs.

Where coded information is to be collected it is essential that the coding categories are decided in detail before the recording begins, as numerous changes in the system during the fieldwork can lead to inconsistencies. The potential data can be readily assessed, so this should not present any major problems. The coding system should allow, however, for additional categories missed during the preliminary evaluation; where there is group recording, all members need to be informed of new categories.

Whilst most data may be easily described and given a code, it is easier for recorders to have images of monument forms on their code sheet. Whilst some can be described with appropriate architectural or other terms, these are often clumsy and obscure to all but the most carefully prepared field worker. Depending on the complexity of the coding, the illustrations may cover one or more page of code sheet.

5.2 Images

Memorials are highly visual artifacts and often require recording in some way. This may be through photography, sketch or measured drawings, or rubbings.

These may record all carved surfaces, or only particular features, and may require the equipment listed in Table 3.

5.2.1 Photography

The photography of graveyard memorials has been considered by a number of researchers (Farber, 1976; Gray and Ferguson, 1997) and the images can be of a quality which are works of art in their own right (Duval and Rigby, 1978). For some images, the need to provide a clean and even colour (removing the distractions of the geology of the stone and any lichen growths) has led to the taking of casts and the photographing of these, though at very considerable cost in time and materials (Rigby with Noordsij, 1992). High quality black and white prints may be necessary for publication. For most purposes, a record photograph is sufficient, with rapidly processed colour 35 mm prints or digital images now the norm. Photographs may be of the whole stone (with complex memorials having images for each face), or of details such as a form of script or decorative motif. Most record photographs should be face-on to the stone, and unwanted vegetation should be cleared away. Digital images should be renamed to indicate site name and memorial number, so that the relevant pictures can be easily and rapidly located at a later date. All digital images should also be copied so that there are archive backup sets so that should the primary record, such as on the lap top hard drive, become corrupted there is still an accessible set of images. Sometimes oblique photographs show detail more effectively that face-on images, as they can show the monument as a three-dimensional artefact, and can emphasise carving of text and decoration. Where appropriate these should also be taken and incorporated into the record.

In common with most archaeological photographs, a scale and board with site code and identifying number is desirable, though these may be omitted for some types of publication photograph. It is vital that the correct number is used in the photograph; the record forms or at least an annotated sketch map should be used to ensure this. A checklist of stones photographed should be kept, or they can be marked off on the map. If memorials do not have any identifying number on the print, then linking images to written record forms can be time consuming and a source of error. Also, numbers on the photograph ensure that negatives and digital images have clear identification.

The most important requirement for a successful gravestone photograph is the light. In order to highlight incised text and carved decoration an oblique light source is required. Whilst this may be provided by flash, sunlight is usually preferable. However, as most gravestones in a burial ground face the same way, the sun will only be at a suitably oblique angle for a relatively short time. This time period can be extended by deflecting sunlight using a mirror, a board covered with aluminium foil, or even a large white card. Some memorials may never be in suitable sunlight because of their position against a wall or under trees. Indirect sunlight, flash, or portable studio lights may then be the only way of improving the lighting. In all cases where flash is used, this must be oblique to the stone, and so needs to be on an extension from the camera. Where flash is only on the camera it may be better to take oblique photographs of the stone. Photography at night can allow complete control over the lighting of a memorial, but local communities and authorities need to be informed, otherwise the activity may be misinterpreted.

5.2.2 *Drawings*

Sketch drawings can be valuable reminders of particular details, particularly where photography is problematic. Features such as methods of construction on complex monuments may be noted with a sketch. More detailed measured drawings may be desirable in some cases, and architectural mouldings can be reproduced using a profile gauge. Measured drawings of plans, elevations and cross sections of complex monuments such as tombs and mausolea may be produced, treating each monument as a small structure.

5.2.3 *Rubbings*

Rubbings, like photographs, can be of the highest quality, and some scholarly publications have relied heavily on these rather than photographs (Kelly and Williams, 1984; Tashjian and Tashjian, 1974). They require considerable skill and patience to reach this standard, though classes are often held at the Association for Gravestone Studies Annual Conference. Lower quality rubbings can still be sufficient for producing a record (figure 52), though it may be necessary to trace over the rubbing in pencil with it laid out in front of the stone to ensure that the necessary detail has been transferred to the paper. Line drawings can be derived from these rubbings and used for further study and publication (Mytum, 2000). Whilst some stone types such as slate produce excellent rubbings, others will be less clear because of the texture of the stone, and for some rock types rubbings would be highly damaging to the surface (Lie, 1976). Care should therefore be taken in evaluating the impact of rubbing before a program is undertaken. The necessary equipment is listed in Table 3.

6. ANALYSIS

For many studies it is crucial to ascertain the likely date for the erection of the memorial, and so this needs to be recorded. This may or may not exactly match the death dates for those listed (Mytum, 2002b). When analysis is to be carried out

Figure 52. Making a rubbing of a headstone using wax crayon and lining wallpaper held in place with elastic ribbon.

by decade, or perhaps quarter century, then it is worth including this at the data entry stage rather than having to carry out various sorting procedures. Databases may be based on memorials, or on individuals commemorated, and this can greatly affect the size of the database.

Many forms of study will involve intuitive examination of images to identify stylistic trends, or the work of particular carvers. During analysis in such cases, the record forms with photographs are usefully placed in folders by decade, so that trends can be observed, and roughly contemporary memorials can be compared. Reduced tracings of rubbings can be similarly organised. Digital images can be selected and arranged on screen to allow visual comparison. Indeed, if a relational database has been constructed, linked to the images, those with particular coded values can be selected and compared relatively easily. Final archive storage should be by memorial number, though concordances produced by computer can provide valuable indices of attribute states.

Much of the analysis will be tailored to the particular research questions and the cultural context of the work, as discussed in previous chapters. Research needs to remain focused, and should not be side-tracked into descriptive accounts of the memorials, or a version of anecdotal local history. On the other hand, the material forms and the local historical context will be important in understanding the evidence.

7. EXCAVATION

Excavation of human remains is an activity which has traditionally been a feature of archaeological research but which has recently been constrained in many parts of the world through pressure from particular ethnic, religious or other interest groups. This has resulted in more prescriptive legislation at national, regional or ecclesiastical levels. The effects of the NAGPRA legislation in the United States have been the most intensively studied (recent bibliographies are available on the World-Wide Web). In many other countries either the government heritage bodies or archaeological professional organisations have now produced guidelines and codes of practice linked to legislation and ethics. Examples include Scotland (Historic Scotland, 1997) and England (Garratt-Frost et al., 1992), where there is also Faculty jurisdiction by the ecclesiastical courts for Church of England burial grounds (Bianco, 1996). Archaeologists have recently become aware of the ethics of excavating the bodies of the more recent past, and there is no consensus on this matter (Boyle, 1999; Cox ,1998b; Reeve and Cox, 1999). Archaeological investigation in relation to forensic science is a developing subject (Hunter et al., 1996; Hunter et al., 1994; papers in *Historical Archaeology* 35.1(2001)).

It is not only legislation that affects the excavation of human remains. The feelings of contemporary local communities, and descendants or other parties who feel that they have an interest in a particular cemetery, have to be considered (Mytum, 2002c). Whilst several high profile cases often attract attention, such as the African Burial Ground, New York City (Harrington, 1996; Kaufman, 1996; LaRoche and Blakey, 1997), in many cases effective liaison locally allows work to proceed with no conflict, but these are rarely mentioned in the literature (though see as examples Faulkner, 2000:31–32 in England, and Goldstein, 1995 in California). Poirer and Bellantoni (1997b) outline a partnership approach involving various interested parties, a policy also explicitly reported by Roberts and McCarthy (1995).

Some archaeologists themselves feel that excavation of human remains is inappropriate, though in situations where the remains are to be disturbed by development or natural erosion, then archaeological excavation may be the most appropriate and respectful way of lifting the remains (figure 53, left). Alternatives such as professional exhumation contractors are often largely motivated by profit and may lead to insensitive treatment of the remains (Reeve and Cox, 1999: 162). The result is that many professional archaeological bodies have produced codes of conduct and guidelines for the excavation, study and reburial of human remains, each reflecting their current legislation and the cultural values of the wider population (Mayes, 1998; Ubelaker, 1989). It is vital that the appropriate national and local legislation is fully understood at the beginning of any project planning, and is fully adhered to throughout. Archaeologists may have to recognise, however, that little if any analysis may be possible during or following excavation, and reburial

Figure 53. Left, Large scale excavation within a medieval and later church prior to major underpinning in advance of mining beneath it, Kellington, North Yorkshire, England. Right, Skeleton within a wooden six-sided coffin, Kellington, North Yorkshire, England.

may happen almost immediately, or within a tightly defined timescale. This will therefore require such on-site anatomical analysis as can be provided; this may also be required where the bone is poorly preserved due to soil conditions. In some cases it is possible to arrange for human remains to be placed in a location which is accessible for further study but also is acceptable to the local community (Mytum, 1997). Where the acidity of the soil is such that no human remains survive, it may be possible to identify the position of bodies through soil phosphate tests.

Artefactual and pathological study may also be restricted if health and safety concerns apply, and bodies that are particularly well preserved may have associated health risks (Reeve and Adams, 1993:17–20). The examination of crypts has led to the development of procedures which protect staff from both physical (Kneller, 1998; Konofes and McGee, 1996) and psychological (Thompson, 1998) risks. This has largely built upon the experience first gained at Spitalfields, London (Reeve and Adams, 1993). It is still unclear what taphonomic processes any corpse may have experienced within a sealed coffin, so on opening such a container a wide range of preservational conditions should be anticipated. Very different remains have been found in adjacent coffins in the same crypt, though in earth graves there seems to be greater predictability (Boddington et al., 1987; Nawrocki, 1995). The frozen remains of the Franklin expedition created other opportunities and dilemmas (Beattie, 1999; Beattie and Geiger, 1993)

An evaluation exercise involving stripping of the topsoil, followed by careful cleaning of the surface (Owsley et al., 1997), can often define the grave cuts, though in some subsoils these can still be difficult to recognize. Graves cut into a natural subsoil and rapidly backfilled with the same clean material can be extremely difficult to identify even for experienced excavators. Small family burial grounds are often located a small distance from rural colonial farmsteads (Gibb, 1996). Those excavating domestic sites should be aware of such a possibility; where burials are found, a sufficient area around known burials should be investigated, as there are often several clusters, each perhaps representing a generation, a particular ownership over several generations, or status differences such as owners and slaves.

Orientation of inhumations is a traditional concern in archaeology, but may be of particular relevance in historic burial. This should be measured with care using a defined and uniformly applied measure, given in degrees; simple cardinal directions are not sufficient. Careful photography with the north point marked, and ideally the planning of the skeletal remains within the grave with north again indicated, allows later reconsideration of orientation. Vertical photography of skeletons can aid the planning process. Digital photography, image processing on a laptop computer, and then the printing out of an image can form the basis of a more detailed annotated drawing of the grave. It is important to differentiate between deliberate positioning of the body within the grave during interment and post-depositional movement during the process of decay. Likewise, the exact location of all artefacts within the grave fill should be carefully recorded, to differentiate between those placed in the grave at interment from others which were merely residual within the grave fill. Of particular significance may be pins indicating a shroud or clothing, and coffin furniture. If the coffin survives, or there are stains, the shape and size of the coffin should be ascertained if at all possible (figure 53, right).

Excavation studies within the historic period have tended to concentrate on demography, health, ethnicity and status, largely based on skeletal analysis (Garland and Janaway, 1989; Grauer, 1995; Poirier and Bellantoni, 1997). There have yet to be many large scale analyses of coffins and their fittings. There has been some interest in the taphonomy associated with burial, both in earth graves (Boddington et al., 1987; Nawrocki, 1995) and in crypts and vaults (Adams and Reeves, 1987; Reeves and Adams, 1993). The linking of historic archaeological data on individuals and communities with that from documentary sources can be particularly rewarding, and can have important implications at a wider methodological level (Molleson and Cox ,1993; Saunders et al., 1995).

There has yet been little opportunity to unite below ground excavated data with that derived from above ground inscribed memorials on any scale. Given the limited nature of excavation on most sites with memorials this is likely to continue. However, the pressures for the clearance of urban cemeteries for reuse or redevelopment, and the unavoidable relocation of some rural burial grounds in the face of large scale engineering schemes, could offer some exciting and important opportunities of which cultural resource managers should be fully aware.

8. CONCLUSIONS

Carrying out graveyard fieldwork can be extremely rewarding, but it can rapidly produce a vast amount of data of very great complexity. There is also the need for very considerable data checking, though this is relatively easy with memorials. With excavation data, checking measures within the excavation and recording system are vital to ensure the quality of the data (or at least to measure the likely levels of error). Well planned fieldwork in burial grounds can be a very efficient way of studying the past, and it can also have a positive contemporary community effect, revitalising interest in local heritage, improving protection of the cultural resource, and increasing a sense of identity for those in the vicinity. Working in graveyards involves awareness of local sensitivities and cultural values, but is also largely appreciated by those with an attachment to the burial place. Publication, deposition of records, and public outreach are therefore important because of the range of people interested in the data and results of fieldwork, and these issues are discussed in the following chapter.

Chapter 10

Conservation, Education, and Display

Many issues of conservation, education and display that apply to burial grounds are similar to those in other aspects of archaeology, and only some of the issues, which are especially significant or may be different in this context, are elaborated upon here. Particular sensitivity is required because of the ethical and emotional factors associated with burial, and greater sensitivity and consultation than is legally required may be appropriate. Archaeologists showing care and respect for burials and monuments can, however, be much appreciated by descendants and local inhabitants alike. Indeed, one of the greatest threats to burial grounds is vandalism, and the archaeologists' interest in the site, if publicized and popularized within the wider community, can do more to ensure its preservation than any professional heritage management strategy alone. That is why conservation, education and interpretation are so closely inter-linked. Many government agencies have produced conservation and management guidelines, some mainly for the public (Fitzpatrick, 1995), others with more detail to support professional or volunteer action (Anson-Cartright, 1997; Maxwell et al., 2001; McGahee, 1997; Thompson, 1989).

1. ETHICS AND METHODS OF CONSERVATION

An essential text for all those involved with conservation issues is that of Strangstad (1988), and members of the Association for Gravestone Studies are involved in many initiatives, some of which are reported in their newsletter. Those who intend to become involved with any conservation programs would be advised to contact the Association, and so reach those most active and experienced in this field. As full recording at all stages is an essential part of the documentation associated with conservation procedures, see Chapter 9 for further details of graveyard recording methods.

The ownership and rights to intervene on memorials and burial plots needs to be ascertained before any work can take place. In Britain, for example, monuments are owned by the heirs of the person who purchased the plot or memorial, not the burial ground owners (Harte, 1994). The issuing of public notices is usually necessary prior to any work, both in churchyards (Stapleton and Burman, 1988) and cemeteries (Dunk and Rugg, 1994), though some work can be undertaken quickly if public safety is involved. In each country, state, or individual burial ground, the laws and regulations may be different and should be carefully checked with the appropriate authorities. The guidelines produced by Lavelle (1992), though designed for New South Wales, contain useful widely applicable advice set within the context of the Burra charter.

1.1 Landscape Conservation

By far the greatest human cause of degradation of the heritage within burial grounds is through inappropriate land management schemes. These can vary from removal of historic tree planting or kerbstones around graves to laying flat of memorials. Even insensitive mowing and use of a nylon whip (weed-eater or strimmer) can do much damage to stones. In many burial grounds, naturally sown trees begin to grow up and disturb monuments through their roots and also grow around memorials. Large scale clearance of memorials for ease of mowing has been a popular solution in many burial grounds, often supported by arguments concerning safety (Dunk and Rugg, 1994). Further proposals for such schemes should be resisted where possible, as they have often been unsuccessful in improving the overall environment, despite their severe degradation of the historic and archaeological evidence. Many small urban burial grounds have been turned into open spaces for leisure (Mytum et al., 1994), often involving movement of memorials; any further schemes should be carefully evaluated to preserve the maximum amount of surviving material evidence.

In all burial areas, from small family graveyards to massive urban cemeteries, vegetation will quickly take hold if an active management strategy is not in place (Thompson, 1989: 37–38). In abandoned burial grounds, and sectors of large cemeteries where not only burial but also grave visiting has largely ceased, the spread of undergrowth and then tree saplings soon hides the monuments (figure 54). As such secondary woodland develops it acquires its own ecological dynamic, which can be considered a valuable resource in itself. In Britain, for example, the environmental lobby has been so successful in publicizing the ecological value of churchyards and cemeteries (Barker, 1977; Dennis, 1993; Freed, 2001; Greenoak, 1985) that if conservation and churchyards are mentioned together, the public assumes this is ecological rather than cultural conservation. The range of plants growing within a churchyard or cemetery can be impressive (Latham, 1984; 1989), and this then supports invertebrate life which, combined with the vegetation and relative tranquillity, in turn encourages a range of reptile, bird (Goodfellow, 1994) and mammal species. Sometimes the monuments themselves are important, with rare lichens growing on some types of rock, and bats roosting in mausolea.

In the long term much damage can be inflicted on monuments by vegetative regeneration. Roots grow around gravestones or cause them to tip over, ivy and

Figure 54. Secondary woodland spreading over the cemetery, Prague, Czech Republic.

other climbing plants enter cracks and pull monuments apart, and branches or whole trees fall during storms and smash even well preserved memorials. There thus has to be a balance between cultural and natural conservation, and it is often necessary to undertake a detailed survey of the memorials at a time of the year when the vegetation has died down so that its cultural value can be appreciated. There can be real conflicts between cultural and ecological conservation, and archaeologists need to gather their evidence and arguments carefully to ensure that their side of the negotiations is given due attention.

1.2 Memorials

One of the most common aspects of conservation necessary in a graveyard or cemetery is that of dealing with uneven ground and leaning monuments due to subsidence. On burial, a coffin takes up a considerable amount of space, but as it decays and collapses, the soil sinks. This can often leave depressions in the ground, and where several coffins have been inserted at the same time or in quick succession, this can lead to substantial collapses. Whilst most of the subsidence is directly over the coffin, movement of soil around the edges of the grave can undermine monuments, causing them to lean.

The re-erection of headstones and other monuments needs to be undertaken with care, particularly with softer stones that may already be suffering erosion.

Careful examination of the monument for weathering and cracks must be undertaken to see if the attempt at moving will in fact lead to its disintegration. Only sound memorials should be re-erected. There are also health and safety considerations, as many stones are larger than at first appear, with quite an amount buried beneath the surface. In many cases leaning stones can be straightened by removing earth from one side and carefully pushing and pulling the stone back into the vertical. If they have to be completely removed and the ground solidified, the task is much more complex and expensive. It is then worth placing them on a buried unmortared brick base, with brick or stone rubble used at the rear to hold the stone in place (Strangstad, 1988: 65–69). Where further subsidence is feared, it may be useful to set the memorial on a base, or even within some form of mortar support, though this may be unsuitable for some stones. Some types of memorial with edging kerbs and the central area covered with concrete or chippings often suffer badly from compaction as coffins and interments decay. Substantial subsidence can lead to the appearance of large voids once the surface has been broken, though the infilling of the void and the resurfacing with concrete is relatively simple.

Large, freestanding, memorials that have become unstable may need to be dismantled, the ground beneath prepared and a platform or raft constructed, and the structure re-erected on that. Many tall pedestal monuments, crosses and statuary need to be assessed and preventative measures taken in advance of collapse, as the damage to the sculpted elements can be considerable if they fall. Major monuments and mausolea in many respects are akin to buildings and can be treated as such. Clearly the costs of stabilization and restoration for complex monuments can be considerable, and should be undertaken within a brief provided by a conservation architect.

Many memorials suffer collapse either from erosion, an accident such as a tree blowing down in a gale, or from vandalism. They have often been repaired and altered in the course of their use lives, but this has not always been sympathetically done, and may lead to further problems at a later date. For example, headstones can split in a variety of ways, related to the structure of the rock out which they are cut. Various methods of using metal clamps have been tried in the past, both in original construction and in repair (figure 55). Where possible on ancient stones, iron should be replaced with stainless steel or another inert, non-corrosive metal. Broken headstones can be mended with a weak mortar mix; too much cement can, as with pointing of historic building fabric or tomb bases, create further problems.

Whilst the rock source chosen for memorials might have been in part selected because of its durability, many were made from the local stone, and others from materials that could be easily and finely carved. Over the course of time, many memorials have weathered to some degree, not helped in many cases by atmospheric pollution. The extent of pollution over time has indeed been measured by examining memorials (Cooke et al., 1995; Livingston and Baer, 1990; Meierding, 1993). Moreover, the combination of materials originally used in composite monuments

Figure 55. Headstone with iron fixings which have corroded and led to the slate splitting, Newport, Pembrokeshire, Wales.

can lead to problems due to differential absorption of water (Robinson, 2001:284). If graveyard memorials are to be considered for conservation, then a detailed fieldwork assessment has to be made. Local conservation offices may have recommended assessment procedures for that region, such as those produced for Scotland (Maxwell et al., 2001; Yates et al., 1998). Many stones will not be conserved because of the cost, and because techniques will not eradicate inherent problems such as damp rising from the ground. Some experiments such as patching stones or coating them with sealant have been shown to do greater harm than leaving the monuments untouched (Hosley and Holcombe, 1994; Strangstad, 1988).

Most conservation projects are not published, but an excellent account of the various stages undertaken for the brownstone memorials at Hartford, Connecticut has been produced (Hosley and Holcombe, 1994). The decision to consolidate and re-carve some of the memorials is controversial, and many conservation programs would not include such work on ethical as well as cost grounds. The program of preservation and restoration of Boston's many historic burial grounds began in 1983 (Chase, 1990). An historic assessment was made of each burial ground (Atwood, 1990; Pollan, 1990a; 1990b) and an extensive program of monument conservation, burial ground reinstatement and landscaping, and overall environmental improvement has been undertaken. Some work has also been carried out on the curation of broken gravestones in New England (Erickson, 1995). In England, some

Figure 56. Restored tomb, Painswick, Gloucestershire, England. Note the well clipped yew trees in the background.

conservation has been achieved within cemeteries, and also a few churchyards, notably Painswick, Gloucestershire (figure 56). The New Orleans monuments have benefitted from the interest of several groups, and their distinctive local character and value as a tourist attraction has combined to allow fundraising for extensive reconstruction and conservation (figure 57, left).

Memorials that will continue to deteriorate severely may be either left in place with protection, moved, or allowed to decay. Glass cases have been used in several North American cemeteries to protect particularly fine carvings. The most famous of these are in Chicago (Brown, 1994: 161–163, 179) and Forest Hills, Boston (Wilson, 1998: 114–115, 122–123) for fine figure sculpture. Such structures preserve the carving *in situ*, but destroy the context and original integrity of the monument. Seeing the decay of nearby uncovered memorials indicates the success in preservation, but at a cost. Some memorials could be moved inside for preservation, but then the purpose of grave marking and the overall landscape setting is lost.

Many of the fine early Scottish sandstone memorials in churchyards are suffering from erosion, but it has been decided that moving or covering them would destroy the integrity of most of the monuments and their settings (Maxwell et al., 2001). They have been given as ideal conditions *in situ* as can be arranged and left to decay, a fitting and appropriate memorial to not only the person but

Figure 57. Left, Rebuilding and restoring a family tomb, St. Louis no. 1 cemetery, New Orleans, Louisiana. Right, Information sign, St. Louis no. 1 cemetery, New Orleans, Louisiana.

also the time in which they were created. If nothing else, mortuary archaeology reminds us that nothing will last for ever.

Casts can be made of particularly significant designs, though they are slow to make, require considerable skill and experience, and then take up much room in storage; one method is described by Rigby with Noordsij (1992). Rubbings and photographs are also important within the context of any conservation or recording program, and have been discussed under fieldwork in Chapter 9. Cleaning of memorials is possible in some circumstances, but professional advice should be obtained. For many rock types only water should be used, with a soft brush or sponge, and where detergents can be used, these are not domestic types (Strangstad, 1988: 60–63).

2. HISTORIC BURIAL GROUNDS IN EDUCATION

Burial grounds are excellent outdoor classrooms where a wide range of elements within the school curriculum can be explored, often in an integrated way, which helps pupils. It is also possible to use such a resource for many ages,

Figure 58. School children working on a graveyard project, Granary burial ground, Boston, Massachusetts.

examining different issues, or similar topics at varying levels of complexity and sophistication (figure 58). The role of graveyards and education has been a common theme amongst a stream of publications, beginning in the 1970s (Duca, 1974). Some later publications integrate the role of computers (Mytum, 2000: 159–162), but otherwise largely consider the same types of themes (Morris and Corbishley, 1996; Purkis, 1993). Education is an issue addressed at the Association for Gravestone Studies conference each year, and for the latest activity in this field those involved with that event should be contacted.

Excavations during conservation are usually only short term, and often undertaken at speed. There can also be legal controls or conditions laid down by those giving permission, which may prevent or severely curtail wider access, particularly where human remains are being exposed. If local schools can visit, however, much can be learnt by observing the skilled, careful and scientific process, and becoming aware of the ethical framework under which archaeologists work. This can greatly improve the image of archaeology, and help bring what can be a disruptive operation for the community into perspective (Crist and Roberts, 1996). Careful control of school parties and appropriate briefing of teachers is essential if such visits are to be effective. In the case of college students, field schools may provide a chance to work on a burial ground excavation and explore first hand the many practical and ethical issues involved in such research linked to conservation or reburial issues. In most cases, however, education can and should be conducted at the vast majority of sites where invasive examination is not necessary.

Graveyard and cemetery surveys provide an ideal opportunity for data collection and analysis in a range of subject areas, such as anthropology, archaeology, architectural and art history, environmental science, geology, history, religion, sociology. This can be carried out examining one site, or at a more comparative, regional level. Some of the issues that could be addressed, and methods to collect the data, have been outlined in previous chapters, but scholars coming from different disciplines will wish to define the questions and methodologies within their own disciplinary traditions. The non-destructive nature of graveyard survey, and its avoidance of many ethical issues otherwise associated with many aspects of the study of mortuary behavior, makes it extremely attractive and viable in contemporary contexts. Many issues raised through graveyard survey can be further examined through documentary study (either directly relevant sources such as cemetery archives or burial registers, or more generalized sources such as census returns and trade directories). Oral history may be appropriate, but sensitivity needs to be paid to this aspect of research.

Fortunately, burial grounds of any period and size are suitable for educational use, but for any programs to be successful, they need to address the detailed curriculum in the education authority responsible. In all countries, teachers are hard pressed to fulfill their obligations under set criteria, and liberal yet unfocussed projects no longer have a place in the timetable. It is therefore only possible here to outline some of the aspects of graveyards that can be used; detailed proposals produced by or for teachers will have to be set in the local context and linked to key subjects such as literacy, mathematics and science.

As with all field visits, issues of health and safety need to be carefully addressed both before and during the visit. Plant and animal life can be dangerous or harmful, depending on the region. Common problems could be poison ivy, brambles and stinging nettles, or even snakes. The safety of trees, burial ground walls and memorials needs to be considered. Students should not be allowed to climb on any monuments for their own safety, and that of the memorials.

The management of a class also needs careful consideration. As it is quite easy for children to be hidden behind stones whilst observing or making notes, it is vital that a sufficient number of assistants are present, and that children work in groups rather than alone. Graveyard recording can anyway be both more fun and more accurate in a collaborative structure. The sensitive context of a burial ground makes some forms of behavior inappropriate, so control has to be kept over noise levels. Students need to be made aware of the significance of flowers and small items placed on graves, and that these should not be touched.

Burial grounds may be one of the few categories of open space in an urban environment, and a rare example of old grassland in intensively farmed regions. They can allow a wide range of ecological issues to be addressed, such as plant diversity, lichens and pollution, bird behavior including nesting. It is the cultural resources available, however, that will be the focus here.

The memorials and their setting can be studied in their own right, as products of artists and craftsmen, or as indications of how people wished to be remembered in the past. The inscriptions provide specific information about past individuals, such as names, ages, dates of death, place of residence. All this can inform about the past, but it is also possible for teachers to use the burial ground merely as an outdoor repository of data that can be collected, classified and analyzed, or utilized for inspiration. Some of the main subject areas are considered below.

2.1 Mathematics

Gravestones provide many opportunities for mathematical projects. They can be counted, in total or by types, and they contain information about many individuals. The information can be considered to show a range of values (years of death, ages at death), averages of various kinds, standard deviations. Sampling is easy within a graveyard, and different sampling methods can be compared and contrasted. Memorials are also easy to measure, to varying levels of accuracy. They can also be mapped, which can incorporate exercises in trigonometry.

2.2 Science

The materials of memorials can provide insight into geology and weathering. Only the earliest graveyards will have limited a range of local stone; most will have a wide array with different surface treatments, and incorporating other materials such as metals, terra-cotta, wood and glass. The various characteristics of different materials can thus be compared. Graveyards may also be suitable habitats for the study of ecology and the biological sciences.

2.3 Information Technology

Graveyard recording data is ideal for considering the design and formation of databases. All students can enter up some records, either their own forms or those of someone else. Problems of data quality and absence of data can arise. Many of the mathematical problems can be addressed with computers, and graphical output of all kinds can be produced. It is easy to think of questions to ask of the data, and then modify the question slightly to see what change there is in the answer. For example, average age of death can be modified to that of males or females, excluding all those under 16, and so on. Whilst these can be investigated manually, they can all be answered very rapidly by computer. Images can be scanned, and CAD used for individual memorials or for burial ground plans. With the wide availability of digital cameras, the integration of photographs with other data is now much easier; CDs can be produced easily for storage and distribution, and can combine interpretative text, a database and images.

2.4 Social History

The individuals listed on particular stones, within burial plots, and in the burial ground as a whole give considerable opportunity to consider family structure, and any changes over time. Demography can be considered, though inscriptions are often not a close reflection of actual patterns of death at the times which are more accurately reflected in burial registers. This contrast in the picture produced by two different sources can be enlightening, and can raise issues such as class and desires for memorialization. Gender can be studied in terms of how people are defined. If the burial ground has representatives of different ethnic groups, they can be compared. If stones are signed by the masons who carved them, the products of different workshops can be can be considered. Most memorials are not signed, but the general changes in style can be compared and linked to overall trends in the economy, including increased use of technology and importation of exotic materials because of improved transport.

2.5 Religion

The degree to which memorials reflect religious beliefs can be examined through the form and motifs on memorials, and through the inscriptions. For single denomination burial grounds changes can be considered over time, and in cemeteries with many denominations and faiths, wider comparisons can be drawn. The relationship of graveyard symbolism to that found in religious books and architecture can be explored. The issue of death and the afterlife can be raised and discussed in a context which is both real and not immediately threatening. The evidence of death at all ages can be useful, as can the ways in which graves are tended, and people mourn for their lost friends and relatives.

2.6 Art

Monuments may rarely be high art, but they represent a major collection of popular culture, and often relate to art styles in fashion at the time of the memorial's construction. The main revival styles are widely represented on headstones as well as large mausolea, as are art nouveau, art deco, and the arts and crafts movements. Ethnic art styles may also be represented. Students can develop their artistic abilities through drawing memorials, or creating new designs either in a particular genre or of their own making. Gravestone rubbing can also be a stimulating exercise on memorials which are resilient enough. Technical drawing can also be incorporated through careful measuring and drawing of monuments, particularly those with many components. New designs can also be drawn out as if for a mason to follow.

2.7 Literature

The texts inscribed on memorials may occasionally be of some quality, though usually the verses are mundane or merely doggerel. They can still be used for critique and an encouragement for poetry, whatever their quality. Creative writing can be inspired by empathy with the deceased, or the lives and relationships of the named individuals from the past. Past and present emotions felt in the overall graveyard setting, or brought on by individual monuments, can also be encouraged. Burial grounds of all kinds always evoke some sort of response that should inspire even the least enthusiastic writer. Visits in different seasons can induce varying responses.

2.8 Folklife Studies

Folklife and folklore allows the memorials and their production to be placed in a wider community context. This may involve stories from the parents and grandparents of students, reminiscences from local monumental masons and funeral directors, and from those who manage the burial grounds. This allows many groups within the community to share their experiences and attitudes with students, and can create links to other forms of memorials away from the burial ground such as photographs, letters, diaries, heirlooms and funerary costume.

2.9 Citizenship

The public obligations in maintaining a burial ground can be a good context in which to understand different viewpoints and obligations. The effects (both physical and emotional) of vandalism can be considered, and the ways in which community support creates a positive environment and sense of identity can be explored.

3. PUBLIC INTERPRETATION AND DISPLAY

The decision as to whether to have any public interpretation of excavations at burial sites depends on the local context. The excavation site may be fenced off so that viewing is deliberately prevented, but information can still be provided on panels for passers by. Structured visits may be allowed, either for particular groups or for the public at set times, but this can be both expensive to organize and easily result in misunderstandings regarding the treatment of remains. Plenty of time must be allowed for observation and questions to be asked. In many circumstances, illustrated lectures during and after the project can be very well received. The importance of communicating with local interested parties the nature of the work and its results is important in all archaeology, but probably never more than with regard to human remains. There may be concerns over security, however, if the excavations cannot be left locked up at night. Disturbance may occur by those hoping for grave goods or wishing to acquire items for occult purposes, and the timing of publicity and visits needs to be structured and programmed with due regard to such threats.

Most burial areas are not being disturbed by archaeologists, but the monuments and their landscape context are still often very deserving of interpretation at some level. Interpretation strategies in cemeteries and burial grounds will vary greatly depending on several key factors. The first is the degree of continuing burial and memorial visitation, and the potential reaction of the bereaved to visitors with other interests. Usually, visitors are welcomed, but sections with only recent graves may well be omitted from interpretation. The second factor is the context of the burial ground, its safety, ease of access, and types of visitor that could be anticipated. Some areas will have a large tourist potential, others will be the local population drawn from a large city, in others a limited number of local people and the occasional passer by. The third factor is the size and nature of the burial ground, as this will affect the content of the interpretation. In overall planning the first two factors are most important. They set the practical constraints within which decisions regarding the third will be made.

Research undertaken by individuals and local history groups may be the basis from which guides can be written. Whilst most has been written about the major cemeteries in large cities (Curl, 2001; Florence, 1997; Linden-Ward, 1989a; Weston, 1989), similar research could be conducted at smaller centers (Murray, 1991; Sledge, 2002); where this has been done, public interest has been considerable. Academic studies may provide the raw materials for public interpretation, but so can the output from student projects, with suitable checking of quality. The literature must be rewritten for the level of prior knowledge expected by the visitors, and jargon of all kinds should be eliminated. The use of photographs can be a great benefit, but often simple line drawings can emphasize the features of most importance, whether it be monument shape, symbols, or lettering style. For all material to be read on site, each guide entry or caption must be relatively brief. For an interpretative program to be most effective, however, there should be an overall theme, which runs through most if not all the text. The types of theme are discussed later, but it is important also to consider the medium of communication to visitors.

The medium for interpretation can be through the use of signs, printed guides of various types from single sheets to books, and guided tours. Each medium has its advantages, but also weaknesses, and more than one can be used in any case.

Signs provide a permanent, easily accessible method of communication, but they are relatively expensive and few cemeteries have many such signs (figure 57, right). There is often a plan provided at major cemeteries so that relatives can find their way to the appropriate plot, but this rarely gives other information or any heritage interpretation, yet this would be easy to incorporate. Key graves and monuments can be indicated on a sign with a suitably clear map, together with a limited amount of historical, art historical, archaeological and biographical information. Rarely have more than one or two signs relating to the heritage been provided in a cemetery. There is, however, no reason why many historic cemeteries should not be treated as any other historic site or landscape, with a series of panels placed at intervals. These could be positioned beside the paths not only at the entrance, but also near particular memorials or zones and at vistas where a range of memorials and the landscape setting could be appreciated.

The self-guided trail based on a leaflet is by far the most common existing form of cemetery interpretation. Key memorials are identified on a map, and in some cases sketches are provided to aid identification on the ground. Passing reference can be made to other features, but a leaflet is limited in the amount of information that can be provided. Its advantage is that it is easy to use, and relatively inexpensive to research and produce. Problems can arise with making leaflets available. If there is a nearby shop or a cemetery office, these are the best outlets, with signs at the cemetery gates explaining where the leaflets may be obtained. An alternative is to make leaflets available for free; a box for financial contributions could be provided, but in some areas this may be too easily vandalized and the contents taken.

Guidebooks have the advantage of allowing considerable detail to be provided, together with more illustrations, but have the disadvantage of cost and relatively limited sales at all but the most popular cemeteries. They, too, could be sold from shops and offices, but could not normally be available for free distribution. A balance has to be struck between length and detail given and costs and time taken to read the guide. It may be best if a longer book is planned, to have some clearly labeled sections with brief on-site guide entries for visitors to use whilst at the site, and other sections that can be read at another time. Repeat visitors can subsequently be more fully informed and so notice and enjoy more aspects of the burial ground. It is vital that the guide is not difficult to use on site; most guide books are often hardly consulted at all at the location, but may perhaps be read or at least looked through later.

Guided tours are always popular at cemeteries, but they suffer from being labor-intensive. The result is that tours tend to be held rarely, and most of those who would appreciate and benefit from a tour are casual independent visitors who could come at any time. If many different guides are employed, there is always the danger of uneven quality of tour, and for many subjects of interest that could be explored in the cemetery, staff or volunteers may be insufficiently knowledgeable

Figure 59. Historically informed tour, La Fayette cemetery, New Orleans, Louisiana.

to give interpretations and answer questions. For example, in New Orleans there are many commercial and some voluntary group tours, but the quality of the historical information is extremely variable (figure 59).

The use of tours, both at single sites, and over a whole or half-day around several burial grounds, can be very successful, particularly in less populated areas. Such events will be of value to local historical societies but can be of general interest, especially if a substantial part of the population has lived in the area for several generations. Tours can be important events at which local interest in cemetery conservation and management can be awakened, and publicity gained through local newspapers, radio and even television.

Burial grounds lend themselves to one of a limited number of themes with regard to interpretation, and these are worth briefly outlining, with examples of the implementation of each.

As open spaces, with deliberate plantings of specimen trees and even flower bedding, burial grounds can be appreciated as a form of park or garden. Besides the plantings provided by the burial ground authorities as part of their management of the area, individual burial plots might have been planted by relatives, reflecting different cultural and aesthetic priorities. The landscaping of 19th-century cemeteries was often to provide a suitable environment not only for burial and mourning, but also promenading and education. In this sense, the modern interpretation of such sites can relive this early purpose. The landscaping and plantings offer a means by which both botanical and cultural themes can be united, thus widening interest to more than just heritage tourists. Given that planting fashions have changed over time, different parts of a cemetery may reveal these chronological trends.

Many burial grounds have become neglected and overgrown. In some cases at least, limited clearance is possible in conservation programs, but rarely can such sites be returned to their former pristine state. Highgate cemetery in London has used the overgrown appearance of much of the original cemetery to emphasize the romantic, with accompanied tours allowing the visitor fleeting glimpses of memorials poking above lush vegetation between the trees (Barker, 1984). The shady setting gives a melancholy and romantic air which resonates with Victorian sentimentality, though of course it is ahistorical in that it is not what the Victorians themselves experienced on their visits to Highgate.

The value of overgrown burial grounds for wildlife, particularly in urban contexts, often gives them a value higher than that attributed to the cultural content, as already discussed under conservation. Interpretation here can consider the whole history of the cemetery and the region it served, showing how areas have gone out of use and reverted to secondary woodland. This may be linked to indigenous trees, or be due to the introductions brought into the cemetery as part of its design. Many small colonial family burial grounds are also now overgrown, and here too the changing appearance of the landscape is worthy of comment.

The majority of tours and the self-guided literature tends to be rather conservative in what is covered, with a particular focus on great individuals buried, occasional anecdotes about people and monuments, and recognition of a few key architects or monumental artists. In this style of interpretation it is on the one hand easy to gather snippets of biographical information, but on the other difficult to fully engage the visitor. Whilst a few cemeteries contain monuments to people who are truly remembered today, most are to local businessmen and politicians. Guides tend to note how successful they were, but not link their achievements to anything that may still have an impact on the lives of visitors. If most visitors are to be local, the ways in which the deceased affected the growth and development of the settlements, established still functioning firms or built major public buildings would be of interest. To tourists from further afield, such information is of little if any concern. It is thus vital that the appropriate audience be identified. Moreover, just a series of short biographies means nothing if there is not a theme (such as the rise and fall of the industries in a town, or its physical development) that means that an overall message is provided. To discover that local people of note were buried in the cemetery is hardly a surprise.

Tours only occasionally mention the great mass of memorials in a burial ground, let alone the apparently open areas that contain the unmarked graves. Discussion of these people, some as individuals, others as groups or classes, could greatly enhance whatever theme was chosen for interpretation. How to use more of the less imposing burial monuments in other ways is discussed below.

One of the main research areas relating particularly to cemeteries and to a lesser extent other burial grounds has been the architectural dimension. Cemetery gates and buildings (offices, chapels and crematoria) may be of merit in their own right, but can in any case display styles and indicate the range of functions considered by both the original designers and subsequent managers. Major monuments and mausolea clearly relate to architecture, but so do many of the smaller memorials and fittings such as plot railings, and the overall design of the cemetery can be discussed in terms of landscape fashions. The range of materials used in memorials is also effectively drawn to the attention of the visitor through architecture, as the popularity and use of materials often mirrors that in buildings, whether in a polychrome Gothic mausoleum or a simple concrete grave marker. A guide themed around geology has been produced for the General Cemetery in Sheffield (Kennett, 1995). The architectural dimension can also be easily linked in many cases to issues of style, and this provides more opportunity to explore individuality and meaning.

Style of memorial design has been an interpretative theme in some areas such as New England where carvers of 18th-century stones have been the subject of much research (Hosley and Holcombe, 1994: 58), and in cemeteries where major artists and sculptors have produced monuments. In the case of memorials produced within a tradition of popular culture, whether in early 18th-century New England or Britain, or early 20th-century ethnic burial grounds in America, other themes can be explored. These include the links between producer and consumer, the meanings of symbols, and the role of particular craftsmen (named or anonymous). The reinforcing or contrasting messages of symbols and text can also be highlighted.

The "Battle of the styles" in the 19th century is also an effective theme, by which various revival styles can be identified and contrasted. They are often alluded to in guides but rarely developed as an integrating feature. There is much for the visitor to notice once the stylistic features of the common revivals such as the Classical, Egyptian and Gothic are pointed out. In many memorials there may be more than one tradition represented, together with features which are linked to ethnic identity (Hosley and Holcombe, 1994: 59). The revival styles may be most obvious on the large monuments and mausolea, but many smaller monuments display these styles in their shape and decorative features.

Anthropological issues have rarely been given more than passing mention in interpretative literature, yet many themes would be ideal for linking visitors to the past, contrasting their beliefs, values and material circumstances with those of different groups at various times in the past. The concepts of death and life, and various methods of defining both the deceased and those with whom they are connected, can be drawn out from the stones. An interpretative theme based around providing the skills and basic information necessary for visitors to decode and consider any memorial is a particularly exciting and liberating one. Visitors may read inscriptions and notice anecdotally the frequency of infant death or the ways in which women were defined, even without any interpretation. If some cultural background is given and examples are pointed out, then the visitors can themselves compare and contrast, using grand monuments or the most humble as they chose. Likewise, changes over time and the spread of fashion through society can be highlighted, and compared with modern fashion trends in material culture. A cemetery visit can thus become a wider educational experience than visitors had anticipated, and can also equip them to view other burial grounds with a critical eye, even if there is no interpretation there. Some burial grounds will also appeal to special interest groups; one such example would be sites with many war graves that would attract those fascinated by military history. Special publications or information on the web might be most effective in catering for such segments of the market.

4. CONCLUSIONS

Conservation of burial markers and landscapes, and indeed the below-ground remains, requires consideration of a number of factors which can involve other professions including ecologists, architects and specialist materials conservators. Decisions have to be made in the light of priorities that will relate to local cultural significance of remains, their academic research value, and their educational and interpretative potential. The use of burial areas for education and leisure as well as the interment of human remains makes them integrated into society and more likely to be preserved. When research at any level is conducted on burial grounds, it is important that the results are fed back to the management and local community so that the resource can be appreciated and valued. Indeed, the conservation process can itself be the subject of an interpretive tour (Hosley and Holcombe, 1994: 59–60). In few other aspects of the historic environment can the general public become so concerned over what is perceived as unseemly conduct, yet cemetery research and interpretation offers an opportunity to show respect for a community and its history, and to revive interest in the material remains of that past.

Appendix 1

Examples of Recording Systems

This appendix provides some information on coding and recording forms. This is not comprehensive, but will allow fieldworkers to use some of the experience others have gained in carrying out their fieldwork. Regional terminologies may well be best maintained, though it is important to ensure that what is being described as a particular form is clearly defined.

The Ontario form (Norris, 1988) represents a form where the range of headstone and other types are illustrated could be adapted easily to local circumstances (figure 60). A wider range of types could easily be included. The alternative, as designed by Mytum (2000), is one where the collected data is set out on the form (figure 61) and the codes with illustrations as necessary are on separate sheets. It would be possible to collect limited data for many memorials by having columns of coded data (Figure 62) which would require careful checking but would be very easy for computer input. How many columns, and what types of data, could be chosen for each project. If set out in landscape format, many columns of data can be collected for each memorial.

The advantage of a simpler form is that much data can be collected rapidly; the problems come if there is much variability (and so many are entered into the 'Other' categories), or if a more refined typology is subsequently required. The hierarchical, more detailed coding system allows for combining of types at various levels of analysis, but requires a more complex set of codes. Once the recorders are used to the system, after just a few hours, this does not present great difficulties. The additional investment comes in designing the coding in the first place. Whether this is worthwhile will depend on the research questions or the detail of the archive required.

Cemetery No. .... Marker No. .... Year of interment ....

FORM (Circle one code no.)

a) Vertical slab variants GRID CELL ....

(Note: Include toppled stones)

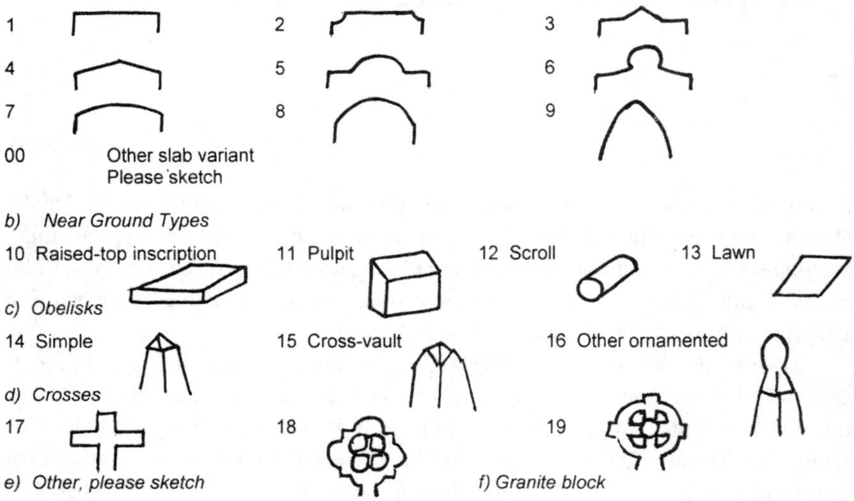

e) Other, please sketch *f) Granite block*

20 21

Lettering is 1 Raised 2 Incised 3 Both

Marker is of 0 D.K. 1 Limestone 2 Granite 3 Slate 4 Sandstone 5 Other ....

Marker orientation 0 N.A. 1 North 2 East 3 South 4 West

Marker height is 0 Less than one 1 2 3 4 5 6 7 8 9+ feet

Motif is 0 Absent 1 Willow 2 Urn 3 Hands clasped 4 Hand pointing up

5 Bible 6 Obelisk 7 Thistle 8 Lamb 9 Other - sketch and/or describe

NOTE: Sculpted motif capping marker (e.g. urn and obelisk) should NOT be recorded.

Marker was supplied by (name) .... of (place) ....

Relationship (h, w, s, d)	*First name*	*Second name*	*Month*	*Day*	*Year*	*Age*	*Place of birth, Country*
....	....	....	....	......	....	....	....
....	....	....	....	......	....	....	....
....	....	....	....	......	....	....	....

NOTE: Enter asterisk if illegible, leave blank if absent

Figure 60. Monument recording form as used in Ontario, Canada (after Norris, 1988)

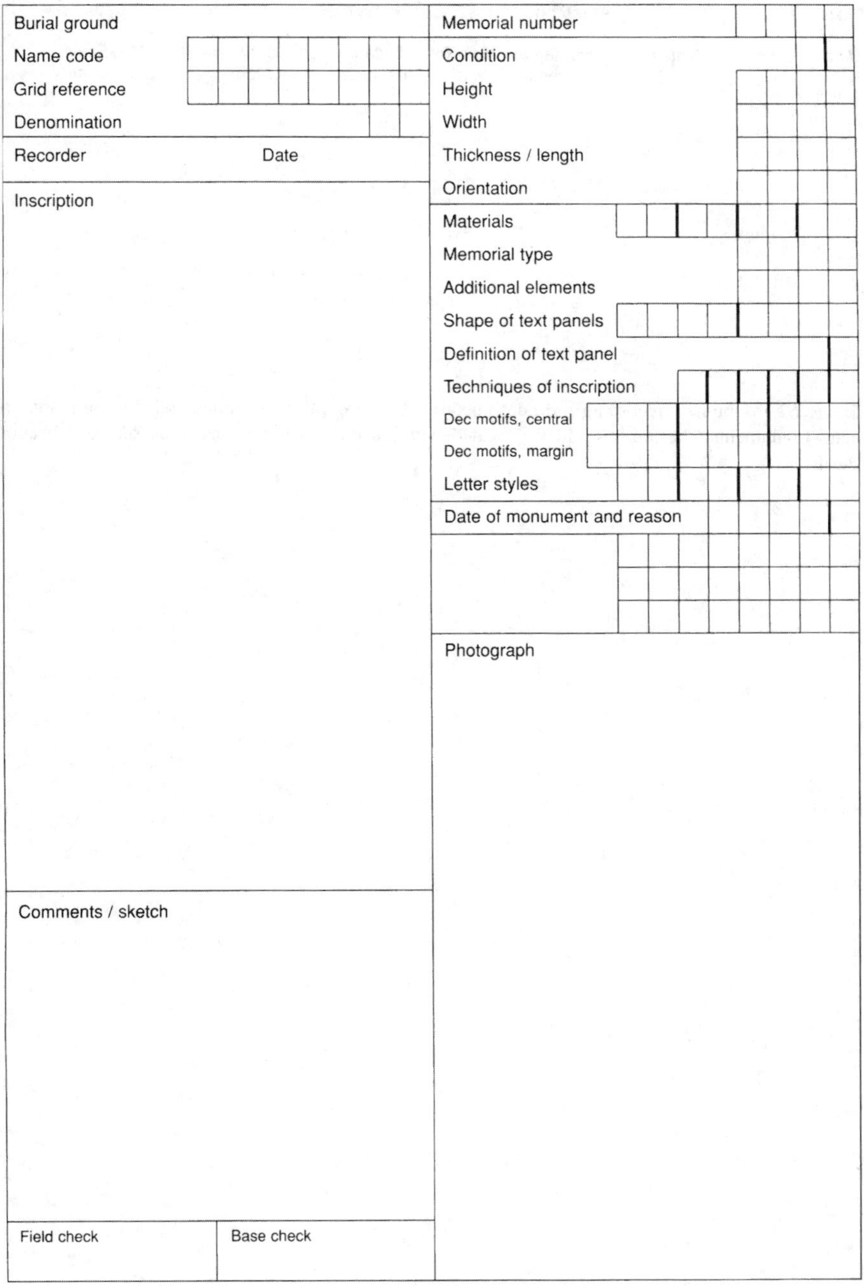

Burial ground
Name code
Grid reference
Denomination
Recorder
Date
Inscription
Comments / sketch
Field check
Base check

Memorial number
Condition
Height
Width
Thickness / length
Orientation
Materials
Memorial type
Additional elements
Shape of text panels
Definition of text panel
Techniques of inscription
Dec motifs, central
Dec motifs, margin
Letter styles
Date of monument and reason
Photograph

Figure 61. Monument recording form using a hierarchical coding system and with space for complete inscription and photograph (Mytum, 2000). The codes would be provided, with illustrations as necessary, on a separate sheet or sheets.

Place .... Area/Zone .... Name of Recorder .... Date ....

Mem. No.	*Form*	*Material*	*Orientation degrees*	*Main Dec motif*	*Lettering technique*	*No. Commem. on stone*	*Date 1st commem*	*Family name*
....	....	....	....	....	....	....	....	....
....	....	....	....	....	....	....	....	....
....	....	....	....	....	....	....	....	....
....	....	....	....	....	....	....	....	....
....	....	....	....	....	....	....	....	....
....	....	....	....	....	....	....	....	....

Figure 62. Monument recording form designed for rapid recording of limited coded information on many monuments. The codes would be provided, with illustrations as necessary, on a separate sheet or sheets.

Appendix 2

Useful Addresses

Association for Gravestone Studies
278 Main Street, Suite 207
Greenfield, MA 01301, USA

Council for British Archaeology
Bowes Morrell House
111 Walmgate
York YO1 9WA, UK

Society for Church Archaeology
c/o Council for British Archaeology
Bowes Morrell House
111 Walmgate
York YO1 9WA, UK

Society for Church Monuments
c/o The Society of Antiquaries of London
Burlington House, Piccadilly
London W1V 0HS, UK

Society for Historical Archaeology
P.O. Box 30446
Tucson, AZ 85751-0446, USA

Bibliography

Adams, M. and Reeve, J., 1987, Excavations at Christ Church, Spitalfields, 1984–6. *Antiquity* 61: 232, 247–256.

Ambler, C., 1998, The New Deal's Landscape Legacy in Kansas Cemeteries. *Markers* 15: 264–285.

Anderson, T.G., 1993, Czech-Catholic Cemeteries in East-Central Texas: Material Culture and Ethnicity in Seven Rural Communities. *Material Culture* 25.3: 1–18.

Anson-Cartright, T., 1997, *Landscape of Memories: A Guide for Conserving Historic Cemeteries, Repairing Tombstones*. Ontario Ministry of Citizenship, Culture and Recreation, Toronto, Ontario.

APT, 1980, *Illustrated Catalogue of American Hardware of the Russell and Erwin Manufacturing Company: An Unabridged Reprint of the 1865 edition*. Association for Preservation Technology, Ottawa.

Ariès, P., 1974, *Western Attitudes toward Death from the Middle Ages to the Present*. Translated by P.M. Ranum. John Hopkins University Press, Baltimore.

—1981, *The Hour of Our Death*. Translated by H. Weaver. Knopf, New York.

—1985, *Images of Man and Death*. Translated by J. Lloyd. Harvard University Press, Cambridge, MA.

Arnold, B. and Wicker, N., 2001, *Gender and the Archaeology of Death*. AltaMira Press, Lanham, MD.

Atwood, R., 1990, Eliot Burying Ground. *Markers* 7: 64–75. (Author's name given in Chase 1990, 62).

Aufderheide, A.C., Neiman, F.C., Wittmers Jr., L.E., and Rapp, B., 1981, Lead in Bone II: Skeletal Lead Content as an Indicator of Lifetime Lead Ingestion and the Social Correlates in an Archeological Population. *American Journal of Physical Anthropology* 55: 285–291.

Bailey, B., 1987, *Churchyards of England and Wales*. Robert Hale, London.

Baird, S., 1992, Language Codes in Texas German Graveyards. *Markers* 9: 216–255.

—1996, The Taylor, Texas, Cemetery: A Language Community. *Markers* 13: 112–141.

Barber, R.J., 1993, The Agua Mansa Cemetery: An Indicator of Ethnic Identification in A Mexican-American Community. In *Ethnicity and the American Cemetery*, edited by R.E. Meyer, pp. 156–172. Bowling Green State University Popular Press, Bowling Green, OH.

Barker, G.M.A., 1977, *Wildlife Conservation in the Care of Churches and Churchyards*. Council for Places of Worship of the General Synod of the Church of England, London.

Barker, F., 1984, *Highgate Cemetery. Victorian Valhalla*. John Murray, London.

Barley, M.W., 1948, Slate headstones in Nottinghamshire. *Thoroton Society Transactions* 52: 69–86.

Barnard, S.M., 1990, *To Prove I'm Not Forgot. Living and Dying in a Victorian City*. Manchester University Press, Manchester.

Bartram, A., 1978, *Tombstone Lettering in the British Isles*. Lund Humphries, London.

Bashford, L., and Pollard, T., 1998, 'In the burying place'—the excavation of a Quaker burial ground. In *Grave concerns: death and burial in England 1700 to 1850*, edited by M. Cox, pp. 154–166. Council for British Archaeology Research Report 113, York.

Beattie, O., 1999, Sleep by the shores of those icy seas: death and resurrection in the last Franklin expedition. In *The Loved Body's Corruption. Archaeological contributions to the study of human mortality*, edited by J. Downes and T. Pollard, pp. 52–68. Cruithne Press, Glasgow.

Beattie, O.B., and Geiger, J.G., 1993, *Frozen in Time: the Fate of the Franklin Expedition*. Bloomsbury, London.

Beier, L.M., 1989, The Good Death in Seventeenth-Century England. In *Death, Ritual and Bereavement*, edited by R. Houlbrooke, pp. 43–61. Routledge, London.

Bell, E.L., 1990, The Historical Archaeology of Mortuary Behavior: Coffin Hardware from Uxbridge, Massachusetts. *Historical Archaeology* 24.3: 54–78.

—1994, *Vestiges of Mortality & Remembrance. A Bibliography on the Historical Archaeology of Cemeteries*. Scarecrow Press, Metuchen, NJ.

—1997, Documentary and comparative scholarly research: context in cemetery archaeology. In *In Remembrance: Archaeology and Death*, edited by D.A. Poirier and N.F. Bellantoni, pp. 219–229. Bergin and Garvey, Westport.

Bellantini, N.F., Sledzik, P.S., and Poirier, D.A., 1997, Rescue, Research, and Reburial: Walton Family Cemetery, Griswold, Connecticut. In *In Remembrance: Archaeology and Death*, edited by D.A. Poirier and N.F. Bellantoni, pp. 131–154. Bergin and Garvey, Westport.

Bender, T., 1974, The Rural Cemetery Movement: Urban Travail and the Appeal of Nature. *New England Quarterly* 47 (June): 196–211.

Benes, P., 1975, Additional Light on Wooden Grave Markers. *Essex Institute Historical Collections* 111: 53–64.

—1976, The Caricature Hypothesis re-examined. In *Puritan Gravestone Art*, edited by P. Benes, pp. 57–67. The Dublin Seminar for New England Folklife, Annual Proceedings, 1976. Boston University Press, Boston.

—1977, *The Masks of Orthodoxy: Folk Gravestone Carving in Plymouth County, Massachusetts, 1689–1805*. University of Massachusetts Press, Amherst.

—1985, A Particular Sense of Doom: Skeletal 'Revivals' in Northern Essex County, Massachusetts, 1737–1784. *Markers* 3: 71–92.

—1987, "Distinguishing Signs of Truly Gracious and Holy Affections": Revival Motifs in Eighteenth-Century New England Gravestone Carvings. In *Mirror and Metaphor. Material and Social Constructions of Reality*, edited by D. Ingersoll and G. Bronitsky, pp. 137–159. University of America Press, Lanham, MD.

Bennett, D.O. 1994, Bury Me in Second Class: Contested Symbols in a Greek Cemetery. *Anthropological Quarterly* 67.3: 122–134.

Bianco, C. 1996, Legislation and Management in England and Wales. In *Church Archaeology. Research directions for the future*, edited by J. Blair and C. Pyrah, pp. 182–188. Council for British Archaeology Research Report 104, York.

Bigger, F.J., 1898, *The Friary of Bun-na-margie*. Belfast, Northern Ireland.

Bigger, F.J. and Hughes, H., 1900, Armorial Sculptured Stones of the County Antrim. *Ulster Journal of Archaeology* 6: 39–35, 90–104, 162–172.

—1901, Armorial Sculptured Stones of the County Antrim. *Ulster Journal of Archaeology* 7: 58–61, 63, 142–157.

—1902, Armorial Sculptured Stones of the County Antrim. *Ulster Journal of Archaeology* 8: 90–94.

—1903, Armorial Sculptured Stones of the County Antrim. *Ulster Journal of Archaeology* 9: 93–96, 131–137.

Blachowicz, J. 1998, The Gravestone Carving Traditions of Plymouth and Cape Cod. *Markers* 15: 38–203.

Blake, C.F. 1993, The Chinese of Valhala: Adaptation and Identity in a Midwestern American Cemetery. *Markers* 10: 52–89.

Blassingame, J. 1972, *The Slave Community: Plantation Life in the Antebellum South*. Oxford University Press, New York.

Boddington, A., Garland, A.N. and Janaway, R.C. eds. 1987, *Death, Decay and Reconstruction: Approaches to Archaeology and Forensic Science*. Manchester University Press, Manchester.

Bollig, M. 1997, Contested Places: Graves and Graveyards in Himba Culture. *Anthropos: International Review of Ethnography and Linguistics* 92. 1–3: 35–50.

Boore, E. 1998, Burial vaults and coffin furniture in the West Country. In *Grave concerns: death and burial in England 1700 to 1850*, edited by M. Cox, pp. 67–84. York, Council for British Archaeology Research Report 113.

Borg, A. 1991, *War Memorials: from antiquity to the present*. Leo Cooper, London.

Bourke, J. 1996, *Dismembering the Male: Men's Bodies, Britain and the Great War*. Reaktion Books, London.

Boyle, A. 1999, A grave disturbance: archaeological perceptions of the recently dead. In *The Loved Body's Corruption. Archaeological contributions to the study of human mortality*, edited by J. Downes and T. Pollard, pp. 187–99. Cruithne Press, Glasgow.

Boyle, A. and Keevill, G. 1998, 'To praise the dead, and anatomie': the analysis of post-medieval burials at St Nicholas, Sevenoaks, Kent. In *Grave concerns: death and burial in England 1700 to 1850*, edited by M. Cox, pp. 85–99. Council for British Archaeology Research Report 113, York.

Brackner, J. 1993, An Overview of the Tombstones of Nineteenth-CenturyAlabama and their Makers. *The Southern Quarterly* 31.2: 21–49.

Bradbury, M. 1999, *Representation of Death: A Social Psychological Perspective*. Routledge, London.

Brady, P. 1993, Florville Foy, F.M.C.: Master Marble Cutter and Tomb Builder. *The Southern Quarterly* 31 (2): 8–20.

Brears, P.C.D. 1981, Heart gravestones in the Calder valley, *Folk Life* 19: 84–93.

Brickley, M., Miles, A. with Stainer, H. 1999, *The Cross Bones Burial Ground, Redcross Way Southwark London. Archaeological Excavations (1991–1998) for the London Underground Limited Jubilee Line Extension Project*. MOLAS Monograph 3. Museum of London Archaeology Service, London.

Briggs, M.W., 1990, Charles Miller Walsh: A Master Carver of Gravestones inVirginia, 1865–1901. *Markers* 7: 139–171.

Broman, E., 2001, Egyptian Revival Funerary Art in Green-Wood Cemetery. *Markers* 18: 30–67.

Bronner, S.J., 2002, Legendary Explanations: The Protection of the Remu Cemetery During the Holocaust. *Markers* 19: 50–63.

Brooke, J.L., 1988, "For Honour and Civil Worship to Any Worthy Person": Burial, Baptism and Community on the Massachusetts Near Frontier, 1730–1790. In *Material Life in America, 1600–1860*, edited by R. Blair St. George, pp. 463–485. Northeastern University Press, Boston, MA.

Brooks, C., 1989, *Mortal Remains: the history and present state of the Victorian and Edwardian cemetery*. Wheaton Publishers, Exeter, England.

Brothwell, D.R., 1981, *Digging up bones*. British Museum, London. 3rd edition.

Brown, A., 1989, The stonemason. In *The Sleeping City. The Story of Rockwood Necropolis*, edited by D.A. Weston, pp. 54–59. Society of Australian Genealogists and Hale and Iremonger, Sydney, NSW.

Brown, F., 1973, *Père la Chaise: Elysium as Real Estate*. Viking, New York.

Brown, I.W. 1992, The Lamson-Carved Gravestones of Watertown, Massachusetts. In *The Art and Mystery of Historical Archaeology: Essays in Honor of James Deetz*, edited by A.E. Yentsch and M.C. Beaudry, pp. 165–191. CRC Press, Boca Raton, FL.

—1993, The New England Cemetery as a Cultural Landscape. In *History from Things. Essays on Material Culture*, edited by S. Lubar and W.D. Kingery, pp. 140–159. Smithsonian Institution Press, Washington, DC.

Brown, J.G., 1994, *Soul in the Stone: cemetery art from America's heartland*. University of Kansas Press, Lawrence.

Bruce, A., 1997, *Monuments, Memorials and the local historian*. Historical Association, London. Historical Association Helps for Students of History 99.

Buckeye, N., 1977, Early American Gravestone Studies: Their Structure of the Literature. In *Puritan Gravestone Art*, edited by P. Benes, pp. 130–136. The Dublin Seminar for New England Folklife, Annual Proceedings, 1976. Boston University Press, Boston.

Buckham, S., 1999, 'The Men that Worked for England They Have Their Graves at Home.' Consumerist issues within the production and purchase of gravestones in Victorian York. In *The familiar past? Archaeologies of Britain 1550–1950*, edited by S. Tarlow and S. West, pp. 199–214. Routledge, London.

Burgess, F., 1963, *English Churchyard Memorials*. Lutterworth Press, London.

Burke, C., and Betteridge, C., 1989, From bush to burial ground: the landscape and design of a garden cemetery. In *The Sleeping City. The Story of Rockwood Necropolis*, edited by D.A. Weston, pp. 61–73. Society of Australian Genealogists and Hale and Iremonger, Sydney, NSW.

Burley, D.V., 1995, Contexts of Meaning: Beer Bottles and Cans in Contemporary Burial Pracices in the Polynesian Kingdom of Tonga. *Historical Archaeology* 29.1: 75–83.

Burnston, S.A., 1997, The Invisible People: The Cemetery at Catoctin Furnace. In *In Remembrance: Archaeology and Death*, edited by D.A. Poirier and N.F. Bellantoni, pp. 93–103. Bergin and Garvey, Westport.

Bury, S., 1985, *An Introduction to Sentimental Jewellery*. H.M.S.O. London.

Butler, L.A.S., 1964, Minor medieval monumental sculpture in the East Midlands. *Archaeological Journal* 16: 111–153.

Butler, L.A.S., and Morris, R., 1994, Derby Cathedral: The Cavendish Vault. *Derbyshire Archaeological Journal* 116: 14–28.

Caldwell, D.H., 1976, A group of post-medieval noble burials at Heddington. *Transactions of the East Lothian Antiquarian, Field and Naturalists Society* 15: 25–37.

Cannadine, D., 1981, War and death, grief and mourning in modern Britain. In *Mirrors of Mortality: Studies in the Social History of Death*, edited by J. Whaley, pp. 187–242. Europa Press, London.

Cannon, A., 1989, The historical dimension in mortuary expressions of status and sentiment. *Current Anthropology* 30.4: 437–458.

Carrott, R.G., 1978, *The Egyptian Revival: Its Sources, Monuments, and Meaning, 1808–1858*. University of California Press, Berkeley.

Caulfield, E., 1967, Connecticut Gravestones XII. John Hartshorne (1650–c. 1738) vs. Joshua Hempstead (1678–1758). *Connecticut Historical Society Bulletin* 32: 65–79. Also reprinted in Caulfield 1991: 165–188.

—1991, Connecticut Gravestone Articles. *Markers* 8: 9–337. A collection of 18 papers reprinted with some additions in this journal, dates of originals 1951–1979/80.

Chase, T., 1990, Boston's Historic Burying Grounds initiative. *Markers* 7: 59–63.

Chase, T., and Gabel, L.K., 1988, Seven Initial Carvers of Boston 1700–1725. *Markers* 5: 210–232. Reprinted in T. Chase and L.K. Gabel 1997, 40–61.

—1997 *Gravestone Chronicles*. 2 volumes. 2nd edition. New England Historic Genealogical Society, Boston.

Chater, A.O., 1976, Early Cardiganshire Gravestones. *Archaeologia Cambrensis* 125: 140–161.

—1977, Early Cardiganshire Gravestones. *Archaeologia Cambrensis* 126: 116–138.

Childe, V.G., 1956, *Piecing Together the Past: the interpretation of archaeological data*. Routledge & Kegan Paul, London, England.

Chow, C., and Teather, E.K., 1998, Chinese Graves and Gravemarkers in Hong King. *Markers* 15: 286–317.

Christison, D. 1902, The Carvings and Inscriptions on the Kirkyard Monuments of the Scottish Lowlands. *Proceedings of the Society of Antiquaries of Scotland* 36: 280–457.

—1905 Additional Notes on the Kirkyard Monuments of the Scottish Lowlands. *Proceedings of the Society of Antiquaries of Scotland* 39: 55–116.

Christopher, A.J., 1995, Segregation and Cemeteries in Port Elizabeth, South Africa. *The Geographical Journal* 161.1: 38–46.

Church, R.A., and Smith, B.M.D., 1966, Competition and monopoly in the coffin furniture industry. *Economic History Review* 19: 612–641.

Clabburn, P., 1998, *Samplers*. Shire Publications, Princes Risborough, England.

Clark, E.W., 1989, The Bigham Carvers of the Carolina Piedmont: Stone Images of an Emerging Sense of American Identity. In *Cemeteries and Gravemarkers. Voices of American Culture*, edited by R.E. Meyer, pp. 31–59. UMI Research, Ann Arbor, MI.

Clark, L., 1987, Gravestones: Reflectors of Ethnicity or Class? In *Consumer Choice in Historical Archaeology*, edited by S.M. Spencer-Wood, pp. 383–395. Plenum Press, New York.

Collins, B.L., 1990, Decoration Day at Higgins Chapel Cemetery in Unicoi County, Tennessee: A Time to Remember and Celebrate Life. *Tennessee Folklore Society Bulletin* 54.3: 82–90.

Columbus Coffin Company, 1882, *Illustrated Catalogue, Columbus Coffin Company, Columbus, Oh.* Published in fascimile microfiche, E.R. McKinstry 1984, *Trade Catalogues at Winterthur*, item 618. Clearwater Publishing, New York.

Colvin, H., 1991, *Architecture and the After-Life*. Yale University Press, London.

Combs, D.W., 1986, *Early Gravestone Art in Georgia and South Carolina*. University of Georgia Press, Athens.

Constant, C., and Waern, R., 1997, The Woodland Cemetery: Toward a Spiritual Landscape. *Journal of the Society of Architectural Historians* 56.3: 359–364.

Cooke, R.V., Inkpen, R.J., and Wiggs, G.F.S., 1995, Using Gravestones to assess Changing Rates of Weathering in the United Kingdom. Earth Surface Processes and Landforms. *The Journal of the British Geomorphological Research Group* 20.6: 531–546.

Cotter, J.L., 1958, *Archeological Excavations at Jamestown Colonial National Historical Park and Jamestown National Historic Site, Virginia*. Archeological Research Series 4. National Park Service, Washington, DC.

Courtaud, P., Delpuech, A., and Romon, T., 1999, Archaeological Investigations at Colonial Cemeteries on Guadaloupe: African Slave Burial Sites or Not? In *African Sites Archaeology in the Caribbean*, edited by J.B. Haviser, pp. 277–290. Ian Randle, Kingston, Jamaica.

Coutu, J., 1997, Carving Histories: British Sculpture in the West Indies. *Church Monuments* 12: 77–85.

Cox, M., 1996, *Life and Death in Spitalfields 1700 to 1850*. Council for British Archaeology, York.

—1998a, *Grave concerns: death and burial in England 1700 to 1850*. Council for British Archaeology Research Report 113, York.

—1998b, Eschatology, burial practice and continuity: a retrospection from Christ Church, Spitalfields. In *Grave concerns: death and burial in England 1700 to 1850*, edited by M. Cox, pp. 112–125. Council for British Archaeology Research Report 113, York.

Cox, M., and Mays, S., editors, 2000, *Human osteology in archaeology and forensic science*. Greenwich Medical Media, London.

Cox, M., and Stock, G., 1995, Nineteenth Century Bath-Stone Walled Graves at Nicholas's Church, Bathampton. *Somerset Archaeological and Natural History Society* 138: 131–150.

Cressy, D., 1997, *Birth, Marriage and Death: Ritual, Religion and the Life-Cycle in Tudor and Stuart England*. Oxford University Press, Oxford.

Crist, T.A.J., and Roberts, D.G., 1996, Engaging the Public Through Mortuary Archaeology at Philadelphia's First African Baptist Church Cemeteries. *Cultural Resource Management* 19.10: 5–7.

Crist, T.A.J., Roberts, D.G., Pitts, R.H., McCarthy, J.P. and Parrington, M., 1997, The First African Baptist Church Cemeteries: African-American Mortality and Trauma in Antebellum Philadelphia. In *In Remembrance. Archaeology and Death*, edited by D.A. Poirier and N.F. Bellantoni, pp. 19–49. Bergin and Garvey, Westport CT.

Crossley, H., 1991, *Lettering in Stone*. Self Publishing Association,Upton-upon-Severn, England.

Crowell, E.A., and Mackie, N.V. III., 1990, The Funerary Monuments and Burial Patterns of Colonial Tidewater Virginia, 1607–1776. *Markers* 7: 103–138.

Cunningham, K., 1989, Navajo, Mormon, Zuni Graves: Navajo, Mormon, Zuni Ways. In *Cemeteries and Gravemarkers. Voices of American Culture*, edited by R.E. Meyer, pp. 197–215. UMI Research, Ann Arbor, MI.

—1993, The People of Rimrock Bury Alfred K. Lorenzo: Tri-Cultural Funerary Practice. In *Ethnicity and the American Cemetery*, edited by R.E. Meyer, pp. 173–192. Bowling Green State University Popular Press, Bowling Green, OH.

Cunnington, P., and Lucas, C., 1972, *Costume for Births, Marriages and Deaths*. Adam and Charles Black, London.

Curl, J.S. 1972, *The Victorian Celebration of Death*. David and Charles, Newton Abbot, England.

—1980, *A Celebration of Death: An Introduction to Some of the Buildings, Monuments, and Settings of Funerary Architecture in the Western European Tradition*. Constable, London.

—1984, The Design of the Early British Cemeteries. *Journal of Garden History* 4.3: 223–253.

—1994, Young's Night Thoughts and the Origins of the Garden Cemetery. *Journal of Garden History* 14: 92–118.

—2001, editor, *Kensal Green Cemetery. The Origins & Development of the General Cemetery of All Souls, Kensall Green, London, 1824–2001*. Phillimore, Chichester, England.

Danforth, L.M., 1982, *The death rituals of rural Greece*. Princeton Universirty Press, Princeton.

Davies, D., 1996, Dirt, death, decay and dissolution: American denial and British avoidance. In *Contemporary Issues in the Sociology of Death, Dying and Disposal*, edited by G. Howarth and P.C. Jupp. Macmillan, Basingstoke, England.

Davies, P., 1986, *Splendours of the Raj: British Architecture in India 1660–1847*. Murray, London.

Deetz, J.F., 1977, *In Small Things Forgotten: The Archaeology of Early American Life*. Anchor Books, Garden City, NY.

—1983, Scientific Humanism and Humanistic Science: a Plea for Paradigmatic Pluralism in Historical Archaeology. *Geoscience and Man* 23: 27–34. Also reprinted in *Mirror and Metaphor. Material and Social Constructions of Reality*, edited by D.W. Ingersoll and G. Bronitsky, pp. 367–379. University Press of America, Lanham, MD.

Deetz, J.F., and Dethlefsen, E., 1965, The Doppler Effect and Archaeology: A Consideration of the Spatial Aspects of Seriation. *Southwest Journal of Anthropology* 21.3: 196–206.

—1967, Death's Head, Cherub, Urn and Willow. *Natural History* 76.3: 29–37.

—1971, Some Social Aspects of New England Colonial Mortuary Art. In *Approaches to the Social Dimensions of Mortuary Practices*, edited by J.A. Brown. pp. 30–38. Society for American Archaeology, Washington, DC.

de hÓir, S., 1997, 'Instruments of the Passion' on some Kilkenny headstones. In *Kilkenny. Studies in Honour of Margaret M. Phelan*, edited by J. Kirwan, pp. 152–158. Kilkenny Archaeological Society, Kilkenny, Ireland.

Dennis, E., 1993, The Living Churchyard—Sanctuaries for Wildlife. *British Wildlife* 4.4: 230–241.

Dethlefsen, E., 1969, Colonial Gravestones and Demography. *American Journal of Physical Anthropology* 31:321–334.

—1981, The Cemetery and Culture Change: Archaeological Focus and Ethnographic Perspective. In *Modern Material Culture: The Archaeology of Us*, edited by R.A. Gould and M.B. Schiffer, pp. 137–159. Academic Press, New York.

—1992, Strange attractors and the Cemetery Set. In *The Art and Mystery of Historical Archaeology: Essays in Honor of James Deetz*, edited by A.E. Yentsch and M.C. Beaudry, pp. 149–164. CRC Press, Boca Raton, FL.

Dethlefsen, E., and Deetz, J.F., 1966, Death's heads, Cherubs and Willow trees: Experimental Archaeology in Colonial Cemeteries. *American Antiquity* 31.4: 502–510.

—1967, Eighteenth Century Cemeteries: A Demographic View. *Historical Archaeology* 1: 40–42.

Dimmick, L., 1992, Thomas Crawford's Monument for Amos Binney in Mount Auburn Cemetery, "A Work of Rare Merit". *Markers* 9: 158–195.

Dobson, M.J., 1997, *Contours of Death and Disease in Early Modern England*. Cambridge University Press, Cambridge.

Dorsey, S.P., 1957, *Early Churches in America 1607–1807*. Oxford University Press, New York.

Douglas, A., 1975, Heaven Our Home: Consolation Literature in the Northeastern United States, 1830–1880. In *Death in America*, edited by D.E. Stannard, pp. 49–68. University of Pennsylvania Press, Philadelphia.

—1977, *The Feminization of American Culture*. Alfred A Knopf, New York.

Douglass, W. 1969 *Death in Murelaga. Funerary Ritual in a Spanish Basque Village*. University of Washington Press, Seattle, WA.

Duca, A.M. 1974, *Journal from the Gloucester Experiment*. New England Program in Teacher Education, Gloucester, MA.

Dunk, J., and Rugg, J., 1994, *The Management of Old Cemetery Land. Now and the Future. A Report of the University of York Cemetery Research Group*. Shaw & Sons, Crayford, England.

Dunn, J., 1988, *The Parramatta Cemeteries. St Patrick's*. Parramatta and District Historical Society, Parramatta, NSW.

—1991, *The Parramatta Cemeteries. St John's*. Parramatta and District Historical Society, Parramatta, NSW.

Duval, F.Y., and Rigby, I.B., 1978, *Early American Gravestone Art in Photographs*. Dover Publications, New York.

Eckert, E., 1993, Language Change: Testimony of Czech Tombstone Inscriptions in Praha, Texas. In *Varieties of Czech: Studies in Czech Sociolinguistics*, edited by E. Eckert, pp. 79–121. Rodopi, Amsterdam, Netherlands.

—1998, Language and Ethnicity Maintenance: Evidence of Czech Tombstone Inscriptions. *Markers* 15: 204–233.

—2001, Gravestones and the Linguistic Ethnography of Czech-Moravians in Texas. *Markers* 18: 146–187.

—2002, From Moravia to Texas: Immigrant acculturation at the Cemetery. *Markers* 19: 174–211.

Edison, C., 1985, Custom-Made Gravestones in Early Salt Lake City: The Work of Four English Stonecarvers. *Utah Historical Quarterly* 56: 310–330.

Ehl, P., et al., 1991, *Old Bohemian and Moravian Jewish Cemeteries*. Paseka, Prague, Czech Republic.

Elia, R.J., 1992, Silent Stones in a Potter's Field: Grave Markers at the Almshouse Burial Ground in Uxbridge, Massachusetts. *Markers* 9: 132–157.

Elia, R.J., and Wesolowsky, A.B., 1989, *Archaeological Excavations at the Uxbridge Almshouse Burial Ground in Uxbridge, Massachusetts*. Office of Public Archaeology, Boston University Report of Investigations 76, Boston.

Elliott, W.R., 1978, 'Chest tombs' and 'tea-caddies' by Cotswold and Severn, *Transactions of the Bristol and Gloucestershire Archaeological Society* 95: 68–85.

Ellwood, B.B., 1990, Electrical Resistivity Surveys in Two Historical Cemeteries in Northeast Texas: A Method for Delineating Unidentified Burial Shafts. *Historical Archaeology* 24.2: 91–98.

Ellwood, B.B., et al., 1994, Search for the Grave of the Hanged Texas Gunfighter, William Preston Longley. *Historical Archaeology* 28.3: 94–112.

Erickson, H.A., 1995, Historic Gravestone Fragments: A Collections Management Plan. *Northeast Historical Archaeology* 24: 9–18.

Erwin, P.F., 1993, Scottish, Irish and Rom Gypsey Funeral Customs and Gravestones in Cincinnati Cemeteries. In *Ethnicity and the American Cemetery*, edited by R.E. Meyer, pp. 104–130. Bowling Green State University Popular Press, Bowling Green, OH.

Etlin, R.A., 1984, *The Architecture of Death: the Transformation of the Cemetery in Eighteenth-century Paris*. MIT Press, Cambridge, MA.

Fairey, W.B., 1990, The Changing York County, South Carolina Tombstone Business, 1750–1850. *Journal of Early Southern Decorative Arts* 16.2: 1–29.

Farber, D., 1976, Photography of Early Gravestone Art. In *Puritan Gravestone Art*, edited by P. Benes, pp. 123–129. The Dublin Seminar for New England Folklife, Annual Proceedings, 1976. Boston University Press, Boston.

Farber, D., and Farber, J.L., 1988, Early Pennsylvania Gravemarkers. *Markers* 5: 96–121.

Farrell, J.J. 1980, *Inventing the American Way of Death, 1830–1920*. Temple University Press, Philadelphia, PA.

Faulkner, N., 2000, Archaeology from below. *Public Archaeology* 1.1: 21–33.

Fellows, F.P., nd (c. 1850), *Untitled catalogue, F.P. Fellows, Wolverhampton, England*. Published in fascimile microfiche, E.R.McKinstry 1984, *Trade Catalogues at Winterthur*, item 1095. Clearwater Publishing, New York.

Fenza, P.J., 1989, Communities of the Dead: Tombstones as a Refection of Social Organization. *Markers* 6: 136–157.

Ferguson, K.E., and Turnbull, P., 1996, Narratives of History, Nature, and Death at the National Memorial Cemetery of the Pacific. *Frontiers: a Journal of Women's Studies* 16.2/3: 1–23.

Fields, J., and Stacpoole, J., 1973, *Victorian Auckland*. John McIndoe, Dunedin, New Zealand.

Finch, J., 2000, *Church Monuments in Norfolk before 1850. An archaeology of commemoration*. British Archaeological Reports British Series 317, Oxford.

Finnell, D.V., 1993, Fifty Years of Reliability: The Stonecarving Career of Charles Lloyd Neale (1800–1866) in Alexandria, Virginia. *Markers* 10: 91–115.

Fitzpatrick, E., editor, 1995, *The Care and Conservation of Graveyards*. The Heritage Service, Dublin, Ireland.

Fitzpatrick, L., 1997, Secular, Savage and Solitary: Death in Australian Painting. In *The Unknown Country: Death in Australia, Britain and the USA*, edited by K. Charmaz, G. Howarth and A. Kellehear, pp. 15–30. Macmillan Press, Basingstoke, England.

Fleming, D.H., 1931, *St Andrews Cathedral Museum*. Oliver and Boyd, Edinburgh, Scotland.

Florence, R., 1997, *New Orleans Cemeteries. Life in the Cities of the Dead*. Batture Press, New Orleans, LA.

Forbes, H.M., 1927, *The Gravestones of Early New England*. Houghton Mifflin, Boston, MA. Reprinted 1955, Pyne Press, Princeton; 1967, Da Capo Press, New York; 1989, Barre Granite Association, Vermont, VA.

—1990, Symbolic Cemetery Gates in New England. *Markers* 7: 3–18. Reprinted from *Old-Time New England* 24.2, 46–58.

Forman, B.M., 1968, New Light on Early Grave Markers. *Essex Institute Historical Collections* 104: 127–129.

Foster, G.S., and Hummel, R.L., 1995, The Adkins-Woodson Cemetery: A Sociological Examination of Cemeteries as Communities. *Markers* 12: 92–117.

Fox Jr., R.A., 1993, *Archaeology, History, and Custer's Last Battle*. University of Oklahoma Press, Norman.

Francaviglia, R.V., 1971, The Cemetery as an Evolving Cultural Landscape. *Annals of the Association of American Geographers* 61.2: 501–509.

Francis, D., Kellaher, L., and Neophytou, G., 2000, Sustaining Cemeteries: The user perspective. *Mortality* 5.1: 34–52.

Freed, T.H., 2001, The Land Use and Flora and Fauna of Kensal Green Cemetery. In *Kensal Green Cemetery. The Origins & Development of the General Cemetery of All Souls, Kensall Green, London, 1824–2001*, edited by J.S. Curl. pp. 297–326. Philimore, Chichester, England.

Freeman, J.A., 1993, The Protestant Cemetery in Florence and Anglo-American Attitudes Toward Italy. *Markers* 10: 218–242.

Fritz, P.S., 1981, From "Public" to "Private": The Royal Funerals in England, 1500–1830. In *Mirrors of Mortality: Studies in the Social History of Death*, edited by J. Whaley, pp. 61–79. Europa Press, London.

—1995, The Undertaking Trade in England: Its Origins and Early Development, 1660–1830. *Eighteenth-Century Studies* 28: 241–253.

Gabel, L.K., 1994, Ritual, Regalia and Remembrance: Fraternal Symbolism and Gravestones. *Markers* 11: 1–27.

—2002, A Common Thread: Needlework Samplers and American Gravestones. *Markers* 19: 18–49.

Garland, A.N., and Janaway, R.C., 1989, The Taphonomy of Inhumation burials. In *Burial Archaeology: Current Research, Methods and Developments*, edited by C.A. Roberts, F. Lee and J. Bintliff, pp. 15–37. British Archaeological Reports British Series 211, Oxford.

Garman, J.C., 1994, Viewing the Color Line Through the Material Culture of Death. *Historical Archaeology* 28.3: 74–93.

Garratt-Frost, S.J., Harrison, G., and Logie, J.G., 1992, *The Law and Burial Archaeology*. Birmingham: Institute of Field Archaeologists Technical Paper 11.

Gayle, M., 1990, A Portfolio of Harriette Forbes's Cast-Iron Gates. *Markers* 7: 19–33.

Genovese, E.D., 1974, *Roll, Jordan, Roll: The World the Slaves Made*. Pantheon, New York.

Gentle, R., and Feild, R., 1975, *English Domestic Brass 1680–1810, and the history of its origins*. Dutton, New York.

Gibb, J.G., 1996, *The Archaeology of Wealth. Consumer Behavior in English America*. Plenum, New York.

Gillon, E.V. Jr., 1972, *Victorian Cemetery Art*. Dover, New York.

Gittings, C., 1984, *Death, burial and the individual in early modern England*. Croom Helm, London.

—1989, Why was death so big in Victorian Britain? In *Death, Ritual and Bereavement*, edited by R. Houlbrooke, pp. 105–117. Routledge, London.

—1999, Sacred and Secular: 1558–1660. In *Death in England. An illustrated history*, edited by P.C. Jupp and C. Gittings, pp. 147–173. Manchester University Press, Manchester.

Goldstein, L., 1995, Politics, Law, Pragmatics, and Human Burial Excavations: An Example from Northern California. In *Bodies of Evidence. Reconstructing History through Skeletal Analysis*, edited by A.L. Grauer, pp. 3–17. Wiley-Lis, New York.

Goodfellow, P., 1994, *Birds in Churchyards*. Church and Conservation Project, Stoneleigh, England.

Goody, J., and Poppi, C., 1994, Flowers and Bones: Approaches to the Dead in Anglo-American and Italian Cemeteries. *Comparative Studies in Society and History* 36.1: 146–175.

Gorman, F., and DiBlasi, M., 1976, Nonchronological Sources of Variation in the Seriation of Gravestone Motifs in the Northeast and Southeast Colonies. In *Puritan Gravestone Art*, edited by P. Benes, pp. 79–87. The Dublin Seminar for New England Folklife, Annual Proceedings, 1976. Boston University Press, Boston.

Gosnell, L., and Gott, S., 1989, San Fernando Cemetery: Decorations of Love and Loss in a Mexican-American Community. In *Cemeteries and Gravemarkers. Voices of American Culture*, edited by R.E. Meyer, pp. 217–236. UMI Research, Ann Arbor, MI.

Gough, P., 1996, Conifers and Commemoration: The Politics and Protocol of Planting. *Landscape Research* 21.1: 73–87.

Gradwohl, D.M., 1993a, The Jewish Cemeteries of Louisville, Kentucky: Mirrors of Historical Processes and Theological Diversity Through 150 Years. *Markers* 10: 116–149.

—1993b, Intra-Group Diversity in Midwest American Jewish Cemeteries: An Ethnoarchaeological Perspective. In *Archaeology of Eastern North America: Papers in Honor of Stephen Williams*, edited by J.B. Stoltman, pp. 363–382. Archaeological Report 25, Mississippi Department of Archives and History, Jackson, MS.

—1997, Cemetery Symbols and Contexts of American Indian Identity: The Grave of Painter and Poet T.C. Cannon. *Markers* 14: 1–33.

—1998 *Benditcha Sea Vuestra Memoria*: Sephardic Jewish Cemeteries in the Caribbean and Eastern North America. *Markers* 15: 1–29.

Graham, A., 1958, Headstones in Post-Reformation Scotland. *Proceedings of the Society of Antiquaries of Scotland* 91: 1–9.

Grauer, A.L., editor, 1995, *Bodies of Evidence. Reconstructing History through Skeletal Analysis*. Wiley-Lis, New York.

Graves, T.E., 1988, Pennsylvania-German Gravestones: An Introduction. *Markers* 5: 60–95.

—1993, Keeping Ukraine Alive through Death: Ukrainian-American Gravestones as Cultural Markers. In *Ethnicity and the American Cemetery*, edited by R.E. Meyer, pp. 36–76. Bowling Green State University Popular Press, Bowling Green, OH.

Gray, T.E., and Ferguson, L.M., 1997, *Photographing Carved Stones. A Practical Guide to Recording Scotland's Past*. Pinkfoot Press, Balgavies, Scotland.

Greenoak, F., 1985, *God's Acre*. Orbis/W.I. Books, London.

Greenwood, D., 1990, *Who's Buried Where in England*. 2nd edition. Constable, London.

Grenham, J., 1992, *Tracing Your Irish Ancestors*. Gill and Macmillan, Dublin, Ireland.

Grogan, E., 1998, Eighteenth century headstones and the stone mason tradition in county Wicklow: the work of Dennis Cullen of Monaseed. *Wicklow Archaeology and History* 1: 41–63.

Habenstein, R.W., and Lamars, W.M., 1981, *The History of American Funeral Directing*. Bulfen, Milwaukee, WI.

Hadley, D.M., 2001, *Death in Medieval England. An Archaeology*. Tempus, Stroud, England.

Hallam, E. and Hockey, J. 2001 *Death, Memory and Material Culture*. Berg, Oxford, England.

Halporn, R., 1993, American Jewish Cemeteries: A Mirror of History. In *Ethnicity and the American Cemetery*, edited by R.E. Meyer, pp. 131–155. Bowling Green State University Popular Press, Bowling Green, OH.

Hammond, N., 1999, Outpost of Empire: Church Monuments in Belize. *Church Monuments* 14: 129–139.

—2000, Beyond the Mexique Bay: Church Monuments in Belize, Part II. *Church Monuments* 15: 89–102.

Handler, J.S., 1997, An African-type Healer/Diviner and His Grave Goods: A Burial from a Plantation Slave Cemetery in Barbados, West Indies. *International Journal of Historical Archaeology* 1.2: 91–130.

Handler, J.S., and Lange, F.W., 1978, *Plantation Slavery in Barbados: An Archeological and Historical Investigation*. Harvard University Press, Cambridge.

Hannon, B., 1990, The Forgetting Rate: Evidence from a Country Cemetery. *Landscape Journal* 9.1: 16–21.

Hannon, T.J. Jr., 1989, Western Pennsylvania Cemeteries in Transition: A Model for Subregional Analysis. In *Cemeteries and Gravemarkers. Voices of American Culture*, edited by R.E. Meyer, pp. 237–257. UMI Research, Ann Arbor, MI.

Harding, C.J., 1987, Post-medieval Coffin Furniture, in *Wharram: A Study of Settlement in the Yorkshire Wolds, III: Wharram Percy: The Church of St Martin*, by R.D. Bell, M.W. Beresford et al., pp. 150–153 and fiche. Society for Medieval Archaeology Archaeological Monograph Series 11, London.

Harding, W.E., 1976, Zerubbabel Collins' Successor and his Work in Bennington County, Vermont. (Includes discussion, notably by Gamage, Ludwig and Slater). In *Puritan Gravestone Art*, edited by P. Benes. pp. 14–22. The Dublin Seminar for New England Folklife, Annual Proceedings, 1976. Boston University Press, Boston.

Hardwick, W.G., Claus, R.J., and Rothwell, D.C., 1971, Cemeteries and urban land value. *The Professional Geographer* 23: 19–21.

Harfield, A., 1984, *Christian Cemeteries and Memorials in Malacca*. British Association for Cemeteries in South Asia, London.

—1985 *Bencoolen. The Christian Cemetery and The Fort Marlborough Monuments*. British Association for Cemeteries in South Asia, London.

Harrington, P.M., 1996, Bones and Bureaucrats: New York's Great Cemetery Imbroglio. In *Archaeological Ethics*, edited by K.D. Vitelli, pp. 221–235. AltaMira Press, Walnut Creek, CA.

Harrison, J.G., 1990, Some Early Gravestones in the Holy Rude Kirkyard, Stirling. *Forth Naturalist and Historian* 13: 79–96.

Harrison, M., 1980, *Victorian Stained Glass*. Barrie and Jenkins, London.

Harte, J.D.C., 1994,. Law after Death, or 'Whose Body Is It?' The Legal Framework for the Disposal and Remembrance of the Dead. In *Ritual and Remembrance. Responses to Death in Human Societies*, edited by J. Davies, pp. 200–237. Sheffield Academic Press, Sheffield, England.

Hawker, R.W., 1990, In the Way of the White Man's Totem Poles: Stone Monuments Among Canada's Tsimshian Indians, 1879–1910. *Markers* 7: 213–231.

—1991, A Faith of Stone: Gravestones, Missionaries, and Culture Change and Continuity among British Columbia's Tsimshian Indians. *Journal of Canadian Studies / Revue d'Etudes Canadiennes* 26.3: 80–100.

Herbert, A., 1944, Swithland slate headstones. *Transactions of the Leicestershire Archaeological Society* 22.3: 215–240.

Hirschorn, A.S., 1973, Mourning Fans. *Antiques* 103: 801–804.

Historic Scotland, 1997, *The treatment of human remains in archaeology*. Historic Scotland, Edinburgh.

Hobbs, J.H., 2002, Say It With Flowers in the Victorian Cemetery. *Markers* 19: 240–271.

Hockey, J., Katz, J., and Small, N., editors, 2001, *Grief, Mourning and Death*. Open University Press, Buckingham, England.

Hodder, I., 1986, *Reading the Past*. Cambridge University Press, Cambridge.

Horton, L.N., 1997, The Remarkable Crosses of Charles Andera. *Markers* 14: 110–133.

Hosley, W., and Holcombe, S.M., 1994, *By Their Markers Ye Shall Know Them. A Chronicle of the History and Restorations of Hartford's Ancient Burying Ground*. The Ancient Burying Ground Association, Hartford, CN.

Houlbrooke, R., 1998, *Death, Religion, and the Family in England, 1480–1750*. Clarendon Press, Oxford.

Howarth, G., 1997, Professionalising the Funeral Industry in England 1700–1960. In *The Changing face of Death. Historical Accounts of Death and Disposal*, edited by P.C. Jupp and G. Howarth, pp. 120–134. Macmillan, Basingstoke, England.

Howarth, G., and Jupp, P., editors, 1997, *The Changing Face of Death: Historical Accounts of Death and Disposal*. St Martin's Press, New York.

Hubbard, W., 1984, A Meaning for Monuments. *The Public Interest* 78: 17–30.

Humphreys, S.C., and King, H., eds., 1981, *Mortality and Immortality: The Anthropology and Archaeology of Death*. Routledge, London.

Hunter, J.R., Heron, C., Janaway, R.C., Martin, A.L., Pollard, A.M. and Roberts, C.A., 1994, Forensic Archaeology in Britain. *Antiquity* 68: 758–769.

Hunter, J.R., Roberts, C.A., and Martin, A.L., editors, 1996, *Studies in Crime: An Introduction to Forensic Archaeology*. Batsford, London.

Huntington, R., and Metcalf, P., 1979, *Celebrations of Death: The Anthropology of Mortuary Ritual*. Cambridge University Press, Cambridge.

Husbands, C., 1980, Gravestones and local history: problems of interpretation. *The Local Historian* 14.3: 149–158.

Ingersoll, D.W., and Nickell, J.N., 1987, The Most Important Monument: The Tomb of the Unknown Soldier. In *Mirror and Metaphor. Material and Social Constructions of Reality*, edited by D.W. Ingersoll and G Bronitsky, pp. 199–225. University Press of America, Lanham, MD.

Inglis, K., 1989, Men, Women and War Memorials: Anzac Australia. In *Learning About Women: Gender, Politics and Power*, edited by J. Conway, S.C. Bouque, and J.W. Scott. University of Michigan Press, Ann Arbor.

—1992a World War I Memorials in Australia. *Guerres mondiales et conflits contemporains* 167: 51–58.

—1992b, "The Homecoming": The War Memorial Movement in Cambridge, England. *Journal of Contemporary History* 27: 583–605.

—1993, Entombing Unknown Soldiers from London and Paris to Baghdad. *History and Memory* 5: 7–31.

—1998, *Sacred Places: War Memorials in the Australian Landscape*. Mieyungah Press, Carlton, Victoria.

Inguanti, J.J., 2000, Domesticating the grave: Italian-American Memorial Practices at New York's Calvary Cemetery. *Markers* 17: 8–31.

Jalland, P., 1996, *Death in the Victorian Family*. Oxford University Press, Oxford.

—1999, Victorian death and its decline: 1850–1918. In *Death in England. An illustrated history*, edited by P.C. Jupp and C. Gittings, pp. 230–255. Manchester University Press, Manchester.

Jamieson, R.W., 1995, Material Culture and Social Death: African-American Burial Practices. *Historical Archaeology* 29.4: 39–58.

Janaway, R., 1993, The Textiles. In *The Spitalfields Project. Volume 1—The Archaeology. Across the Styx*, by J. Reeve and M. Adams, pp. 92–119. Council for British Archaeology Research Report 85, York.

—1998, An introductory guide to textiles from 18th and 19th century burials. In *Grave concerns: death and burial in England 1700 to 1850*, edited by M. Cox, pp. 17–32. Council for British Archaeology Research Report 113, York.

Jeane, D.G., 1969, The Traditional Upland South Cemetery. *Landscape* 18: 39–41.

—1978, The Upland South Cemetery: An American Type. *Journal of Popular Culture* 11: 895–503.

—1987, Rural Southern Gravestones: Sacred Artefacts in the Upland South Folk Cemetery. *Markers* 4: 55–84.

—1989, The Upland Folk Cemetery Complex: Some Suggestions of Origin. In *Cemeteries and Gravemarkers. Voices of American Culture*, edited by R.E. Meyer, pp. 107–136. UMI Research, Ann Arbor, MI.

Jones, C.R., 1980, Ithamar Spauldin, Stonecarver of Concord, Massachusetts. *Markers* 1: 51–55.

Jones, J., 1976, *How to record graveyards*. Rescue, Trust for British Archaeology and the Council for British Archaeology, London. Reprinted 1979, 1984.

Jones, S., 1997, *The Archaeology of Ethnicity. Constructing identities in the past and present*. Routledge, London.

Jordan, T.G., 1982 *Texas Graveyards: A Cultural Legacy*. University of Texas Press, Austin, TX.

Jumonville, F.M., 1993, The Wastebasket and the Grave: Funeralia in the South. *Southern Quarterly* 31.2, 98–118.

Jupp, P.C., and Walter, T., 1999 The healthy society: 1918–1998. In *Death in England. An illustrated history*, edited by P.C. Jupp and C. Gittings, pp. 256–282. Manchester University Press, Manchester.

Kamerman, J.B., 1988, *Death in the Midst of Life. Social and Cultural Influences on Death, Grief and Mourning*. Prentice Hall, Englewood Cliffs, NJ.

Kaufman, E., editor, 1996 *Reclaiming Our Past, Honoring Our Ancestors: New York's Eighteenth Century African American Burial Ground and the Memorial Competition*. African Burial Ground Competition Coalition, New York.

Kay, W.J. et al., editors, 1984, *Pet Loss and Human Bereavement*. Ames, IA.

Keister, D., and Cronin, X., 1997, *Going Out in Style: The Architecture of Eternity*. Facts on File Inc., New York.

Kelly, A., 1990, *Mrs Coade's Stone*. The Self Publishing Association, Upton-upon-Severn, England.

Kelly, S., and Williams, A., 1984, "And the Men who Signed Them": The Signed Gravestones of New England. *Markers* 2: 1–103.

Kelso, W.M., 1992, Big Things Remembered: Anglo-Virginian Houses, Armorial Devices, and the Impact of Common Sense. In *The Art and Mystery of Historical Archaeology: Essays in Honor of James Deetz*, edited by A.E. Yentsch and M.C. Beaudry, pp. 127–145. CRC Press, Boca Raton, FL.

Kemp, B., 1980, *English church monuments*. Batsford, London, England.

Kenna, M.E., 1976, Houses, fields and graves: property and ritual obligation on a Greek island. *Ethnology* 15.1: 21–34.

—1991, The power of the dead: changes in the construction and care of graves and family vaults on a small Greek island. *Journal of Mediterranean Studies* 1.1: 101–119.

Kennett, P., 1995, *The General Cemetery, Sharrow, Sheffield. A Geological Trail*. Sheffield, England.

Khudabux, M.R., 1999, Effects of Life Conditions on the Health of a Negro Slave. In *African Sites Archaeology in the Caribbean*, edited by J.B. Haviser, pp. 291–312. Ian Randle, Kingston, Jamaica.

Kiest, K.S., 1993, Czech Cemeteries in Nebraska from 1868: Cultural Imprints on the Prarie. In *Ethnicity and the American Cemetery*, edited by R.E. Meyer, pp. 77–103. Bowling Green State University Popular Press, Bowling Green, OH.

King, J.A., Bevan, B.W., and Hurry, R.J., 1993, The Reliability of Geophysical Surveys at Historic-Period Cemeteries: An Example from the Plains Cemetery, Mechanicsville, Maryland. *Historical Archaeology* 27.3: 4–16.

King, J.A., and Ubelaker, D., editors, 1996, *Living and Dying on the 17th Century Patuxent Frontier*. Maryland Historical Trust Press, Crownsville, MD.

Kloberdanz, T.J., 1983, Iron Lilies, Eternal Roses: German-Russian Cemetery Folk Art in Perspective. *Journal of the Americal Historical Society of Germans from Russia* 6.3: 27–31.

Kneller, P., 1998, Health and safety in church and funerary archaeology. In *Grave concerns: death and burial in England 1700 to 1850*, edited by M. Cox, pp. 181–189. Council for British Archaeology Research Report 113, York.

Knoblock, G.A., 1996, From Jonathan Hartshorne to Jeremiah Lane: Fifty Years of Gravestone Carving in Coastal New Hampshire. *Markers* 13: 74–111.

Konofes, J.L. and McGee, M.K. 1996 Old Cemeteries, Arsenic, and Health Safety. *Cultural Resource Management* 19.10: 15–18.

Krüger-Kahloula, A., 1989, Tributes in Stone and Lapidiary Lapses: commemorating Black People in Eighteenth- and Nineteenth-Century America. *Markers* 6: 32–100.

—1993, Homage and Hegemony: African-American Grave Inscription and Decoration. In *Slavery in the Americas*, edited by W. Binder, pp. 317–335. Konigshausen and Neumann, Wurzburg, Germany.

Kselman, T.A., 1993, *Death and the Afterlife in Modern France*. Princeton University Press, Princeton.

LaRoche, C.J., and Blakey, M.L., 1997, Seizing Intellectual Power: The Dialogue at the New York African Burial Ground. *Historical Archaeology* 31.3: 84–106.

Latham, J.B., 1984, A Survey of the Flora of Kensal Green and St. Mary's Cemeteries, 1981–1983. *London Naturalist* 63 (November): 53–67.

—1989, Further Studies of Plant Groups in Kensal Green and St. Mary's Cemeteries, 1983–1989. *London Naturalist* 68 (November): 15–24.

Lavelle, S., 1992, *Cemeteries. Guidelines for their Care and Conservation*. Department of Planning, Heritage Council of New South Wales, Sydney.

Leaney, J., 1989, Ashes to ashes: cremation and the celebration of death in nineteenth-century Britain. In. *Death, Ritual and Bereavement*, edited by R. Houlbrooke, pp. 118–135. Routledge, London.

Lees, H., 1993, *Hallowed Ground. Churchyards of Gloucestershire and The Cotswolds*. Thornhill Press, Cheltenham, England.

Levine, G.S., 1978, Colonial Long Island Grave Stones: Trade Network Indicators, 1670–1799. In *Puritan Gravestone Art II*, edited by P. Benes, pp. 46–57. The Dublin Seminar for New England Folklife, 3, 1978. Boston University Press, Boston.

Lie, J., 1976, Stone Rubbings: Are Model Laws Needed? A Seminar Open Forum. In *Puritan Gravestone Art*, edited by P. Benes, pp. 92–105. The Dublin Seminar for New England Folklife, Annual Proceedings, 1976. Boston University Press, Boston.

Linden-Ward, B., 1989a, *Silent City on a Hill. Landscapes of Memory and Boston's Mount Auburn Cemetery*. Ohio State University Press, Columbus.

—1989b, Strange but Genteel Pleasure Grounds: Tourist and Leisure Uses of Nineteenth-Century Rural Cemeteries. In *Cemeteries and Gravemarkers. Voices of American Culture*, edited by R.E. Meyer, pp. 293–328. UMI Research, Ann Arbor, MI.

—1990, "The Fencing Mania": The Rise and Fall of Nineteenth-Century Funerary Enclosures. *Markers* 7: 34–57.

Linden-Ward, B., and Sloane, D.C., 1985, Spring Grove: The Founding of Cincinnati's Rural Cemetery, 1845–1855. *Queen City Heritage* 43.1: 17–32.

Linden-Ward, B., and Ward, A., 1985, Spring Grove: The Role of the Rural Cemetery in American Landscape Design. *Landscape Architecture* 75.5:126–131, 140.

Lindley, K., 1965, *Of Graves and Epitaphs*. Hutchinson, London.

—1972, *Graves and Graveyards*. Kegan Paul, London.

Litten, J., 1985, Post-medieval burial vaults: their construction and contents. *Bulletin of the CBA Churches Committee* 23: 9–17.

—1991, *The English way of death: the common funeral since 1450*. Robert Hale, London.

—1997 the Funeral Trade in Hanoverian England 1714–1760. In *The Changing face of Death. Historical Accounts of Death and Disposal*, edited by P.C. Jupp and G. Howarth. pp. 48–61. Macmillan, Basingstoke, England.

—1998, The English funeral 1700–1850. In *Grave concerns: death and burial in England 1700 to 1850*, edited by M. Cox, pp. 3–16. Council for British Archaeology Research Report 113, York.

—1999, Tombs Fit For Kings: Some Burial Vaults of the English Aristocracy and Landed Gentry of the Period 1650–1850. *Church Monuments* 14: 104–128.

Little, B., Lamphear, K., and Owsley, D., 1992, Mortuary display and status in a nineteenth-century Anglo-American cemetery in Manassas, Virginia. *American Antiquity* 57.3: 397–418.

Little, M.R., 1989, Afro-American Gravemarkers in North Carolina. *Markers* 6: 102–134.

—1998, *Sticks and Stones. Three centuries of North Carolina gravemarkers*. University of North Carolina Press, Chapel Hill.

Livingston, R.A., and Baer, N.S., 1990, Use of Tombstones in Investigations of Deterioration of Stone Monuments. *Environmental Geology and Water Sciences* 16.1: 83–90.

Llewellyn, N., 1991, *The Art of Death. Visual Culture in the English Death Ritual c. 1500–c.1800*. Reaktion Books, London.

—1996, Honour in Life, Death and the Memory: Funeral Monuments in Early Modern England. *Transactions of the Royal Historical Society*, 6th series, 6: 179–200.

—2000, *Funeral Monuments in Post-Reformation England*. Cambridge University Press, Cambridge.

Longfield, A.K., 1943, Some 18th Century Irish Tombstones. I. Introduction and Dennis Cullen of Monaseed. *Journal of the Royal Society of Antiquaries of Ireland* 73: 29–39.

—1944, Some 18th Century Irish Tomb-stones. II. Miles Brien. *Journal of the Royal Society of Antiquaries of Ireland* 74: 63–72.

—1945, Some 18th Century Irish Tombstones. III James Byrne and his School. *Journal of the Royal Society of Antiquaries of Ireland* 75: 76–88.

—1946, Some 18th Century Irish Tomb-stones. IV. J. Butler, Hugh Rogan, David Doyle, etc. *Journal of the Royal Society of Antiquaries of Ireland* 76: 81–88.

—1947, Some Late 18th and early 19th Century Irish Tombstones. V. Subjects not related to the Crucifixion. Saints and Scenes by Kehoe, at St Mullins, and by an unknown carver, at Termonfeckin. *Journal of the Royal Society of Antiquaries of Ireland* 77: 1–4.

—1948, Some Late 18th and early 19th Century Irish Tombstones. VI. East County Louth. *Journal of the Royal Society of Antiquaries of Ireland* 78: 170–174.

—1954, Some 18th Century Irish Tombstones. VII. Clonmel, Kiltoom, Seir Keiran, etc. *Journal of the Royal Society of Antiquaries of Ireland* 84: 173–178.

—1955, Eighteenth Century Irish Tombstones. VIII. M. and T. Hancock and Miles Brien's "Signature Stone". *Journal of the Royal Society of Antiquaries of Ireland* 85: 114–117.

Lord, J., 1986, Patronage and Church Monuments 1660–1794: A Regional Study. *Church Monuments* 1.2: 95–105.

Lucas, J., 1990, Stonecarvers of Monroe County, Indiana, 128–1890. *Markers* 7: 194–212.

Ludwig, A.I., 1966, *Graven Images: New England Stonecarving and its Symbols, 1650–1815*. Wesleyan University Press, Middletown.

Luti, V.F., 2002, *Mallet and Chisel. Gravestone Carvers of Newport, Rhode Island, in the 18th century*. New England Historic Genealogical Society, Boston.

Lynch, G., 1995, Contemporary Gravemarkers of Youths: Milestones of Our Path Through Pain to Joy. *Markers* 12: 144–159.

MacCloskey, M., 1968, *Hallowed Ground: Our National Cemeteries*. R. Rosen Press, New York.

Mackay, M., 1989, Architectural styles and funerary symbolism. In *The Sleeping City. The Story of Rockwood Necropolis*, edited by D.A. Weston, pp. 30–43. Society of Australian Genealogists and Hale and Iremonger, Sydney, NSW.

Mackie, N.V. III., 1988a, Gravestone Procurement in St. Mary's County, 1634–1820. *Maryland Historical Magazine* 83: 229–240.

Maish, A., Rose, J.C., and Marks, M.K., 1997, Cedar Grove: African-American History in Rural Arkansas. In *In Remembrance: Archaeology and Death*, edited by D.A. Poirier and N.F. Bellantoni, pp. 105–117. Bergin and Garvey, Westport, CT.

Małachovicz, E., 1996, *The Landscape and Architecture of the Vilnius Cemeteries*. Warsaw, Centre for the Preservation of Historic Landscape, 5.8.

Malloy, T.A., and Malloy, B., 1992, The Disappearing Shaker Cemetery. *Markers* 9: 256–274.

—1994, Slavery in Colonial Massachusetts as Seen Through Selected Gravestones. *Markers* 11: 112–141.

—1997, Gravemarkers of the Early Congregational Ministers in North Central Massachusetts. *Markers* 14: 34–85.

Manning, C., 1994, *Clonmacnoise*. Stationery Office, Dublin, Ireland.

Mansfield, N., 1995, Class Conflict and Village War Memorials, 1914–24. *Rural History* 6.1: 67–87.

Masson, A.M., 1993, Père La Chaise and New Orleans Cemeteries. *The Southern Quarterly* 31.2: 82–97.

Matturi, J., 1993, Windows in the Garden: Italian-American Memorialization and the American Cemetery. In *Ethnicity and the American Cemetery*, edited by R.E. Meyer, pp. 14–35. Bowling Green State University Popular Press, Bowling Green, OH.

Maxwell, I, Nanda, R., and Urquhart, D., 2001, *Conservation of Historic Graveyards*. Guide for Practitioners 2. Historic Scotland, Edinburgh.

Mayo, J.M., 1988, *War Memorials as Political Landscape: The American Experience and Beyond.* Praeger, New York.

Mays, S., 1998, *The archaeology of human bones*. Routledge, London.

McCormick, F., 1976, A Group of Eighteenth Century Clogher Headstones. *Clogher Record* 9.1: 5–16.

—1979, A Group of Tradesmen's Headstones. *Clogher Record* 10.1: 12–22.

—1983, The symbols of death and the tomb of John Forster in Tydavnet, Co. Monagahan, *Clogher Record* 11.2: 273–286.

McDowell, P., 1989, J.N.B. de Pouilly and French Sources of Revival Style Design in New Orleans Cemetery Architecture. In *Cemeteries and Gravemarkers. Voices of American Culture*, edited by R.E. Meyer, pp. 137–158. UMI Research, Ann Arbor, MI.

McDowell, P., and Meyer, R.E., 1994, *The Revival Styles in American Memorial Art.* Bowling Green State University Popular Press, Bowling Green, OH.

McGahee, S.H., 1997, *South Carolina's Historic Cemeteries: A Preservation Handbook.* South Carolina Department of Archives and History, Columbia, SC.

McGrath, R.L., 1987, Death Italo-American Style: Reflections on Modern Martyrdom. *Markers* 4: 107–113.

McGuire, R.H., 1988, Dialogues with the Dead: Ideology and the Cemetery. In *The Recovery of Meaning*, edited by M.P. Leone and P.B. Potter, pp. 435–480. Smithsonian Institution Press, Washington, DC.

—1991, Building Power in the Cultural Landscape of Broome County, New York 1880 to 1940. In *The Archaeology of Inequality*, edited by R.H. McGHuire and R. Paynter, pp. 102–124. Blackwell, Oxford.

McKillop, H., 1995, Recognizing Children's Graves in Nineteenth-Century Cemeteries: Excavations in St Thomas Anglican Churchyard, Belleville, Ontario, Canada. *Historical Archaeology* 29.2: 77–99.

McManners, J., 1981, *Death and the Enlightenment: Changing Attitudes to Death among Christians and Unbelievers in Eighteenth-Century France*. Clarendon Press, Oxford.

Meara, D., 1983, *Victorian Memorial Brasses*. Routledge & Kegan Paul, London.

—1991, *A.W.N. Pugin and the Revival of Memorial Brasses*. Mansell, London.

Meierding, T.C., 1993, Marble Tombstone Weathering and Air Pollution in North America. *Annals of the Association of American Geographers* 83.4: 568–588.

Meller, H., 1985, *London Cemeteries. An Illustrated Guide and Gazetteeer*. 2nd edition. Avebury, Amersham, England.

Meridan Britannia Company, 1882, *Catalogue of Casket Trimmings*. Published in fascimile microfiche, E.R. McKinstry 1984, *Trade Catalogues at Winterthur*, item 1667. New York, Clearwater Publishing.

Meyer, R.E., editor, 1989a, *Cemeteries and Gravemarkers: Voices of American Culture*. UMI Research Press, Ann Arbor.

—1989b, Images of Logging on Contemporary Pacific Northwest Gravemarkers. In *Cemeteries and Gravemarkers. Voices of American Culture*, edited by R.E. Meyer, pp. 61–85. UMI Research, Ann Arbor, Mich.

Meyer, R.E., editor 1993a, *Ethnicity and the American Cemetery*. Bowling Green State University Popular Press, Bowling Green, OH.

—1993b, The Literature of Necroethnicity in America: An Annotated Bibliography. In *Ethnicity and the American Cemetery*, edited by R.E. Meyer, pp. 222–237. Bowling Green State University Popular Press, Bowling Green, OH.

—1994, 'And Who Have Seen the Wilderness': The End of the Trail on Early Oregon Gravemarkers. *Markers* 11: 186–219.

—2001, Stylistic Variation in the Western Front Battlefield Cemeteries of World War I Combatant Nations. *Markers* 18: 188–253.

Meyer, R.E., and Gradwohl, D.M., 1995, 'Best Damm Dog We Ever Had': Some Folkloristic and Anthropological Observations on San Francisco's Presidio Pet Cemetery. *Markers* 12: 160–205.

Meyers, M.A., 1975, Gates Ajar: Death in Mormon Thought and Practice. In *Death in America*, edited by D.E. Stannard, pp. 112–33. University of Pennsylvania Press, Philadelphia.

Molleson, T., and Cox, M., 1993, *The Spitalfields Project. Volume 2- the Anthropology. The Middling Sort*. Council for British Archaeology Research Report 86, York.

Moore, J., Blaker, C., and Smith, G., 1991, Cherished are the Dead: Changing Social Dimensions in a Kansas Cemetery. *Plains Anthropologist* 36: 67–78.

Moriarty, C., 1995, The Absent Dead and Figurative First World War Memorials. *Transactions of the Ancient Monuments Society* 39: 7–40.

Morley, J., 1971, *Death, Heaven and the Victorians*. Studio Vista, London.

Morris, R.K., 1989, *Churches in the Landscape*. Dent, London.

Morris, R., and Corbishley, M., 1996, *A Teacher's Guide to Churches, Cathedrals and Chapels*. English Heritage, London.

Murray, H., 1991, *This Garden of Death: The History of York Cemetery*. Friends of York Cemetery, York.

Musgrave, E., 1997, Momento Mori: The Function and Meaning of Breton Ossuaries. In *The Changing face of Death. Historical Accounts of Death and Disposal*, edited by P.C. Jupp and G. Howarth, pp. 62–75. Macmillan, Basingstoke.

Muthiah, S., 1990, *Madras. The Gracious City*. Affiliated East-West Press, New Delhi.

Mytum, H.C., 1988, A Newly Discovered Burial Vault in North Dalton Church, North Humberside. *Post-medieval Archaeology* 22: 183–187.

—1989, Public health and private sentiment: the development of cemetery architecture and funerary monuments from the eighteenth century onwards. *World Archaeology* 21: 283–297.

—1993, Death and Identity: strategies in body disposal and memorial at North Front Cemetery, Gibraltar. In *In Search of Cult. Archaeological investigations in honour of Philip Rahtz*, edited by M.O.H. Carver, pp. 187–192. Boydell, Woodbridge

—1994, Language as symbol in churchyard monuments: the use of Welsh in nineteenth- and twentieth-century Pembrokeshire. *World Archaeology* 26.2: 252–267.

—1996, Intrasite patterning and the Temporal Dimension using GIS: the example of Kellington churchyard. *Annalecta Praehistoria Leidensia* 28: 363–367.

—1997, The reinterment of human remains: the Kellington solution. *Church Archaeology* 1: 50–51.

—1999a, Welsh Cultural Identity in Nineteenth-century Pembrokeshire. The pedimented headstone as a graveyard monument. In *The familiar past? Archaeologies of Britain 1550–1950*, edited by S. Tarlow and S. West, pp. 215–230. Routledge, London.

—1999b, The language of death in a bilingual community: nineteenth-century memorials in Newport, Pembrokeshire. In *Archaeology and Language III: Artefacts, languages and texts*, edited by R. Blench and M. Spriggs, pp. 211–230. Routledge, One World Archaeology 34, London.

—2000, *Recording and Analysing Graveyards*. Council for British Archaeology Practical Handbook 15, York.

—2002a, The Nineteenth-Century Reuse of Gravestones at Cherhill. *Wiltshire Archaeological & Natural History Magazine* 95: 27–33.

—2002b, The dating of graveyard memorials: evidence from the stones. *Post-medieval Archaeology* 36: 1–38.

—2002c Bones and the Community. *The Archaeologist* 45 (Summer 2002): 14–15.

—2002d, Graveyard survey in West Ulster. *Church Archaeology* 5/6.

—2002e, A Comparison of Nineteenth and Twentieth century Anglican and Nonconformist Memorials in North Pembrokeshire. *Archaeological Journal* 159: 194–201.

—2003a, Death and Remembrance in the Colonial Context. In *The Archaeology of the British*, edited by S. Lawrence, pp. 156–173. Routledge, London.

—2003b, Local traditions in early eighteenth-century commemoration: the headstone memorials from Balrothery, County Dublin, and their place in the evolution of Irish and British commemorative practice. *Proceedings of the Royal Irish Academy Section C* 103C.

—2003c, Rural burial and remembrance: changing landscapes of commemoration. In *Age of Industrialisation*, edited by D. Baker. Society for Post-medieval Archaeology, Leeds.

—2003d, A long and complex plot: the development of the marked grave space in Ireland. *Church Archaeology*

Mytum, H.C., Dunk, J., and Rugg, J., 1994, Closed urban churchyards in England and Wales: Some survey results. *Post-medieval Archaeology* 28: 111–114.

Mytum, H. and Evans, R. 2002, The evolution of an Irish graveyard during the 18th century: the example of Killeevan, Co. Monaghan, *Journal of Irish Archaeology* 11: 131–146.

Nakagawa, T., 1990, Louisiana Cemeteries as Cultural Artefacts. *Geographical Review of Japan* 63.2: 139–155.

—1992, Gravestone Landscape Evolution of a Japanese Rural Cemetery. *Geographical Bulletin* 34.2: 82–90.

—1994, Louisiana Cemeteries: Manifestations of Regional and Denominational Identity. *Markers* 11: 28–51.

Nash, G., 2000, Pomp and circumstance: archaeology, modernity and the corporatisation of death: early social and political Victorian attitudes towards burial practice. In *Matter, Materiality and Modern Culture*, edited by P.M. Graves-Brown, pp. 112–130. Routledge, London.

Natural Death Centre, 1996, *Woodland Burial & Nature Reserve Burial Grounds in the United Kingdom*. Natural Death Centre, London.

Nawrocki, S.P., 1995, Taphonomic Processes in Historic Cemeteries. In *Bodies of Evidence. Reconstructing History through Skeletal Analysis*, edited by A.L. Grauer, pp. 49–66. Wiley-Lis, New York.

Nielsen, K.H., 1997, From Society to Burial and from Burial to Society? Some modern analogies. In *Burial and Society. The Chronological and Social Analysis of Archaeological Burial Data*, edited by C.K. Jensen and K.H. Nielsen, pp. 103–110. Aarhus University Press, Aarhus, Denmark.

Nigh, R.F., 1997, Under Grave Conditions: African-American Signs of Life and Death in North Florida. *Markers*: 158–189.

Nilsson, S., 1968, *European Architecture in India 1750–1850*. Faber, London, England.

Noel Hume, I., 1992, *Martin's Hundred*. 2nd edition. Victor Golancz, London, England.

Norris, D.A., 1988, Ontario Gravestones. *Markers* 5: 122–149.

Nylander, C. et. al., 1989, *Protestant Cemetery in Rome: the "Parte Antica"*. Unione Internazionale degli Istituti di Archeologia, Storia e Storia dell'Arte, Rome, Italy.

O'Brien, C., and Sweetman, P.D., 1997, *Archaeological Inventory of County Offaly*. Stationery Office, Dublin, Ireland.

Ó Crualaoich, G., 1998, The 'Merry Wake'. In Irish Popular Culture 1650–1850, edited by J.S. Donnelly Jr. and K.A. Miller, pp. 201–221. Irish Academic Press, Dublin, Ireland.

O'Neill, H., 1965, Iron Tombstones and the Pembrokeshire Coal and Iron Company. *Journal of Industrial Archaeology*, 2.4: 158–160.

Orser, C.E. Jr., 1996, *A Historical Archaeology of the Modern World*. Plenum, New York.

Osborne, B.S., 1974, The Cemeteries of the Midland District of Upper Canada: A Note on Mortality in a Frontier Society. *Pioneer America* 6: 46–55.

Owsley, D.W., 1997, Introduction: New perspectives on the Past. In *In Remembrance: Archaeology and Death*, edited by P.A. Poirer and N.F. Bellantoni. pp. 1–16. Greenwood Publishing Group: Westport, CT.

Owsley, D.W., Ellwood, B.B., and Richardson, M.L., 1997, Locating and Excavating Historic Burials. In *In Remembrance: Archaeology and Death*, edited by P.A. Poirer and N.F. Bellantoni, pp. 201–217. Greenwood Publishing Group: Westport, CT.

—1955, *Meaning in the Visual Arts*. Penguin, Harmonsworth.

Parker–Pearson, M., 1982, Mortuary practices, society and ideology: an ethnoarchaeological study. In *Symbolic and Structural Archaeology*, edited by I. Hodder, pp. Cambridge University Press, Cambridge.

Parrington, M., 1987, Cemetery Archaeology in the Urban Environment: A Case Study from Philadelphia. In *Living in Cities: Current Research in Urban Archaeology*, edited by E. Staski, pp. 56–64. Special Publication 5, Society for Historical Archaeology, California, Pennsylvania.

Patterson, N-L., 1976, The Iron Cross and The Tree of Life: German-Alsatian Gravemarkers in Waterloo Region and Bruce County Roman Catholic Cemeteries. *Ontario History* 48.1: 1–16.

—1989, United Above Though Parted Below: The Hand as Symbol on Nineteenth-Century Southwest Ontario Gravestones. *Markers* 6: 180–206.

Pattison, W.D., 1955, The cemeteries of Chicago: a phase of land utilisation. *Annals of the Association of American Geographers* 45: 245–257.

Penny, N., 1977, *Church Monuments in Romantic England*. London, England.

Penny, N.B., 1974, The Commercial Garden Necropolis of the Early Nineteenth Century and Its Critics. *Garden History* 2 (Summer): 61–76.

Petke, S., 1993, A Chronological Survey of the Gravestones Made by Calvin Barber of Simsbury, Connecticut. *Markers* 10: 1–51.

Phillips, S.M., 1997, Recovering Lost Minds: Evidence of Insanity in a Late Nineteenth-Century Almshouse Skeletal Sample. In *In Remembrance: Archaeology and Death*, edited by P.A. Poirer and N.F. Bellantoni, pp. 79–92. Greenwood Publishing Group: Westport, CT.

Pike, M.V., 1980, In Memory Of: Artefacts Relating to Mourning in Nineteenth Century America. *Journal of American Culture* 3.4: 642–659.

Poirier, D.A., and Bellantoni, N.F., editors, 1997a, *In Remembrance: Archaeology and Death*. Greenwood Publishing Group: Westport, CT.

Poirier, D.A., and Bellantoni, N.F., 1997b, In Partnership: Descendants, Friends-of-the-Dead, and Archaeologists. In *In Remembrance: Archaeology and Death*, edited by P.A. Poirer and N.F. Bellantoni, pp. 231–235. Greenwood Publishing Group: Westport, CT.

Pollan, R., 1990a, Dorchester North Burying Ground. *Markers* 7: 76–88. (Author's name given in Chase 1990, 63).

—1990b, Copp's Hill Burying Ground. *Markers* 7: 89–102. (Author's name given in Chase 1990, 63).

Potterton, H., 1975, *Irish Church Monuments 1570–1880*. Ulster Architectural Heritage Society, Belfast.

Prestiano, R., 1983, The Example of D. Aldo Pitassi: evolutionary thought and practice in contemporary memorial design. *Markers* 2: 203–220.

Price, L.W., 1966, Some Results and Implications of a Cemetery Study. *The Professional Geographer* 18.4: 201–207.

Purkis, S., 1995, *A Teacher's Guide to Using Memorials*. English Heritage, London.

Purnell, N.N., 1993, Oriental and Polynesian Cemetery Traditions in the Hawaiian Islands. In *Ethnicity and the American Cemetery*, edited by R.E. Meyer, pp. 193–221. Bowling Green State University Popular Press, Bowling Green, OH.

Quarles, F., 1635, *Emblems Divine and Moral: together with Heiroglyphics of the Life of Man*. London, G.M. Facsimile reproduction, 1991, Scholars' Facsimiles and Reprints, Delmar, NJ.

Quer, C.S., 1989, La Aparición de Una Nueva Tipología Arquitectónica: el Cementerio. *Jornadas de Arte. El Arte en Tiempo de Carlos III* 4: 207–217.

Quigley, C., 1998, *Modern mummies: the preservation of the human body in the twentieth century*. McFarland & Co., London.

Radford, J. and Farrington, S.M., 2001, *Tombs in Tea. Tea garden cemeteries, Sylhet, Bangladesh*. British Association for Cemeteries in South Asia (BACSA), London.

Ragon, M., 1983, *The Space of Death: A Study in Funerary Architecture, Decoration, and Urbanism*. translated by A. Sheridan. University of Virginia Press: Charlottesville, VA.

Rahtz, P.A., and Watts, L., 1983, *Wharram Percy: the memorial stones of the churchyard*. York University Archaeological Publications 1. University of York, York.

Rahtz, S.P.Q., 1987, The Protestant Cemetery in Rome. Interim report. *Opuscula Romana* 16.10, 149–167.

Rauschenberg, B.L., 1990, A Study of Baroque and Gothic-Style Gravestones in Charleston and Its Environs. *Journal of Early Southern Decorative Arts* 26.1: 19–63.

Rawding, C., 1990, The iconography of churches: a case study of landownership and power in nineteenth-century Lincolnshire. *Journal of Historical Geography* 16.2: 157–176.

Read, B., 1982, *Victorian Sculpture*. Yale University Press, New Haven.

Redknapp, M., 1984, Little Ilford, St Mary the Virgin. *London Archaeologist* 1984: 31–37.

Reeve, J., and Adams, M., 1993, *The Spitalfields Project. Volume 1—the Archaeology. Across the Styx*. Council for British Archaeology Research Report 85, York.

Reeve, J., and Cox, M., 1999, Research and our recent ancestors: post-medieval burial grounds. In *The Loved Body's Corruption. Archaeological contributions to the study of human mortality*, edited by J. Downes and T. Pollard, pp. 159–70. Cruithne Press, Glasgow.

Reid, A., 1909, The Churchyard Memorials of Abercorn, Bowden, and Carrington. *Proceedings of the Society of Antiquaries of Scotland* 44: 33–76.

—1915, Sculptured Sarcophegus and Churchyard Memorials at Dalmeny; with Notes on the Churchyards of Edzell, Lethnot, and Stracathro. *Proceedings of the Society of Antiquaries of Scotland* 49: 285–303.

Remnant, G.L., 1962, Jonathan Harner's Terracottas. *Sussex Archaeological Collections* 100: 142–148.

—1964, Jonathan Harner's Terracottas II. *Sussex Archaeological Collections* 102: 52–54.

Reps, J.W., 1965, Cemeteries, Parks, and Suburbs: Picturesque planning in the Romantic Style. In *The Making of Urban America: A History of City Planning in the United States*, edited by J.W. Reps. Princeton University Press: Princeton.

Richardson, R., 1988, *Death, dissection and the destitute*. Pelican, London.

Richmond, M., 1999, Archaeologia Victoriana: the archaeology of the Victorian funeral. In *The Loved Body's Corruption. Archaeological contributions to the study of human mortality*, edited by J. Downes and T. Pollard, pp. 145–158. Cruithne Press, Glasgow.

Ridlen, S.S., 1996, Tree-Stump Tombstones: Traditional Cultural Values and Rustic Funerary Art. *Markers* 13: 44–73.

Rigby, I.B., with Noordsij, K.M., 1992, Recollections of a Collaboration. A Tribute to the Art of Francis Duval. *Markers* 9: 1–21.

Ritchie, J., 1911, An Account of the Watch Houses, Mortsafes and Public Vaults in Aberdeenshire Churchyards, formerly used for the Protection of the Dead from the Resurrectionists. *Proceedings of the Society of Antiquaries of Scotland* 46: 285–326.

—1920, Relics of the Body-snatchers: Supplementary Notes on Mortsafe Tackle, Mortsafes, Watch Houses and Public Vaults, mostly in Aberdeenshire. *Proceedings of the Society of Antiquaries of Scotland* 60: 221–229.

Roberts, C.A., 1984, Analysis of some human femora from a medieval charnel house at Rothwell parish church, Northampton, England. *Ossa* 9–11: 137–147.

Roberts, D.G., and McCarthy, J.P., 1995, Descendant Community Partnering in the Archaeological and Bioanthropological Investigation of African American Skeletal Populations: Two Interrelated Case Studies from Philadelphia. In *Bodies of Evidence. Reconstructing History through Skeletal Analysis*, edited by A.L. Grauer, pp. 19–36. Wiley-Lis, New York.

Roberts, W.E., 1985, Investgating the Tree-stump Tombstone in Indiana. In *American Material Culture and Folklife: a prologue and dialogue*, edited by S.J. Bronner, pp. 135–145. UMI Research Press, Ann Arbor, MI.

—1988, The Sincerest Form of Flattery: Originals and imitations in 'Rustic Monuments' of the Limestone Belt of Indiana. In *Viewpoints on Folklife: Looking at the Overlooked*, edited by W.E. Roberts, pp. 145–161. UMI Research Press, Ann Arbor, MI.

—1990, Notes on the Production of Rustic Monuments in the Limestone Belt of Indiana. *Markers* 7: 172–193.

Robin, R., 1995, 'A Foothold in Europe': The Aesthetics and Politics of American War Cemeteries in Western Europe. *Journal of American Studies* 29.1: 55–72.

Robinson, J.E., 1994, *Geology in the Churchyard*. Living Churchyard Project and Geologists' Association, Stoneleigh, England.

—2001, A Geology of Kensall Green Cemetery. In *Kensal Green Cemetery. The Origins & Development of the General Cemetery of All Souls, Kensall Green, London, 1824–2001*, edited by J.S. Curl, pp. 275–286. Philimore, Chichester.

Rodwell, W., 1981, *The Archaeology of the English Church*. Batsford, London.

Romaine, L.B., 1976, *A Guide to American Trade Catalogues, 1744–1900*. Arno Press, New York.

Rotundo, B., 1974, Mount Auburn Cemetery: A Proper Boston Institution. *Harvard Library Bulletin* 22.3: 268–279.

—1984, Mount Auburn: Fortunate Coincidences and an Ideal Solution. *Journal of Garden History* 4.3: 255–267.

—1989, Monumental Bronze: A Representative American Company. In *Cemeteries and Gravemarkers. Voices of American Culture*, edited by R.E. Meyer, pp. 263–291. UMI Research, Ann Arbor, MI.

—1997, A Modern Gravestone Maker: Some Lessons for Gravestone Historians. *Markers* 14: 86–109.

Ruby, J., 1984, Post-Mortem Portraiture in America. *History of Photography* 8: 201–222.

—1995, *Secure the Shadow: Death and Photography in America*. MIT Press, Cambridge, MA.

Rugg, J., 1998, A new burial form and its meanings: cemetery establishment in the first half of the 19th century. In *Grave concerns: death and burial in England 1700 to 1850*, edited by M. Cox, pp 44–53. Council for British Archaeology Research Report 113, York.

Saumarez Smith, C. 1998, *The Building of Castle Howard*. Pimlico, London.

Sablow, M., 1993, Southern Cemeteries: Places of Presence. *The Southern Quarterly* 31.2: 157–164.

Sadek, V., and Šedinová, J., 1989, *The Old Jewish Cemetery and the Klausen Synagogue*. State Jewish Museum, Prague.

Sanborn, L.S., 1989, Camposantos: Sacred Places of the Southwest. *Markers* 6: 158–179.

Saunders, S.R., and Herring, A., editors, 1995, *Grave Reflections—Portraying the Past through Cemetery Studies*. Canadian Scholars Press, Toronto.

Saunders, S.R., Herring, A., Shawchuk, L.A., and Boyce, G., 1995, The Nineteenth-century Cemetery at St. Thomas' Anglican Church, Belleville: Skeletal remains, Parish Records and Censuses. In *Grave Reflections—Portraying the Past through Cemetery Studies*, edited by S.R. Saunders and A. Herring, A., pp. 93–117. Canadian Scholars Press, Toronto.

Schoemaker, G.H., 1989, The Shift From Artist to Consumer: Changes in Mormon Tombstone Art in Utah. In *The Old Traditional Way of Life: Essays in Honor of Warren E. Roberts*, edited by R.E. Walls and G.H. Schoemaker, pp. 130–145. Bloomington, IN.

—1992, Acculturation and Transformation of Salt Lake Temple Symbols in Mormon Tombstone Art. *Markers* 9: 196–215.

Schorsch, A., 1979, A Key to the Kingdom: The Iconography of the Mourning Picture. *Winterthur Portfolio* 16 (Spring): 41–71.

Scott, D.D., Fox Jr., R.A., Connor, M.A., and Harmon, D., 1989, *Archaeological Perspectives on the Battle of the Little Bighorn*. University of Oklahoma Press, Norman.

Sewter, A.C., 1975, *The Stained Glass of William Morris and his Circle*. Yale University Press, New Haven, CT.

Shepardson, A.F. 1996, John Huntington, Gravestone Carver of Lebanon, Connecticut. *Markers* 13: 142–222.

Sherlock, P.D., 1999, Academic commemoration: Monuments at Corpus Christi College Oxford 1517–1700. *Church Monuments* 14: 80–87.

Shoesmith, R., 1980, Llangar Church. *Archaeologia Cambrensis* 119: 64–132.

Silvester, B., and Pitman, L., 1998, Eighteenth-century stonemasons in the Black Mountains. *Church Archaeology* 2: 29–34.

Singleton, C.C., and Kay, J., 1989, The Rockwood Cemetery Line. In *The Sleeping City. The Story of Rockwood Necropolis*, edited by D.A. Weston, pp. 44–53. Society of Australian Genealogists and Hale and Iremonger, Sydney, NSW.

Slack, P., 1985, *The Impact of Plague in Tudor and Stuart England*. Oxford University Press, Oxford, England.

Slater, J.A., 1976, Principles and Methods for the Study of the Work of Individual Carvers. In *Puritan Gravestone Art*, edited by P. Benes, pp. 9–13. The Dublin Seminar for New England Folklife, Annual Proceedings, 1976. Boston University Press, Boston.

—1996a, *The Colonial Burying Grounds of Eastern Connecticut and the Men Who Made Them*. Memoirs of the Connecticut Academy of Arts and Sciences. Memoir 21. Revised edition.

—1996b, Jotham Warren, The Plainfield Trumpeter. *Markers* 13: 1–43.

Slater, J.A., and Tucker, R.L., 1978, The Colonial Gravestone Carvings of John Hartshorne. In *Puritan Gravestone Art II*, edited by P. Benes, pp. 79–146. The Dublin Seminar for New England Folklife, 3, 1978. Boston University Press, Boston.

Sledge, J., 1993, A Circular Necropolis on the Gulf Coast: Mobile's Old Catholic Cemetery. *The Southern Quarterly* 31.2: 74–81.

—2002, *Cities of Silence. A Guide to Mobile's Historic Cemeteries*. University of Alabama Press, Tuscaloosa.

Sledzic, P.S., and Hunt, D.R., 1997, Disaster and Relief Efforts at the Hardin Cemetery. In *In Remembrance: Archaeology and Death*, edited by P.A. Poirer and N.F. Bellantoni, pp. 185–198. Greenwood Publishing Group: Westport, CT.

Sloane, D.S., 1991, *The Last Great Necessity. Cemeteries in American History*. Johns Hopkins University Press, Baltimore.

Smith, D.A., 1987, "Safe in the Arms of Jesus": Consolation on Delaware Children's Gravestones, 1840–99. *Markers* 4: 85–106.

—1993, Humble Stones: A Study of Four Monumental Masons. *Ulster Folklife* 39: 1–8.

—1994, Monumental Ambition: A Kentucky Stonecarver's Career. *Markers* 11: 168–185.

Snyder, E.M., 1989, Innocents in a Worldly World: Victorian Children's Gravemarkers. In *Cemeteries and Gravemarkers. Voices of American Culture*, edited by R.E. Meyer, pp. 11–29. UMI Research, Ann Arbor, MI.

Spiegelman, V., and Kastenbaum, R., 1990, Pet Rest Memorial: Is Eternity Running Out of Time? *Omega: Journal of Death and Dying* 21.1: 8–9.

Stanier, D., 1985, The Granite Quarrying Industry in Devon and Cornwall. Part One. 1800–1910. *Industrial Archaeology Review* 7.2: 171–189.

Stannard, D.E., 1973, Death and Dying in Puritan New England. *American Historical Review* 78: 1305–1330.

—1977, *The Puritan Way of Death: A Study in Religion, Culture and Social Change*. Oxford University Press, Oxford.

Stapleton, H., and Burman, P., 1988, *The Churchyards Handbook. Advice on their Care and Maintenance*. 3rd edn. CIO Publishing, London.

Stewart, F., 1978, Life and death in an Oxfordshire churchyard. *The Local Historian* 14.3: 149–158.

Stier, M.M., 1983, 'Wonderfully Lettered and Carved': The Gravestones of the Risley Family, 1786–1835. *Dartmouth College Library Bulletin* 23 new series 2. Hanover, New Hampshire.

Stilgoe, J.R., 1978, Folklore and Graveyard Design. *Landscape* 22.3: 22–28.

Stock, G., 1998a, Quaker burial: doctrine and practice. In *Grave concerns: death and burial in England 1700 to 1850*, edited by M. Cox, pp. 129–143. Council for British Archaeology Research Report 113, York.

—1998b, The 18th and early 19th century Quaker burial ground at Bathford, Bath and North-East Somerset. In *Grave concerns: death and burial in England 1700 to 1850*, edited by M. Cox, pp. 144–153. Council for British Archaeology Research Report 113, York.

Stocker, B., 1986, Medieval Grave Markers in Kent. *Church Monuments* 1.2: 106–114.

Stokes, S., 1991, Gone But Not Forgotten: Wakulla County's Folk Graveyards. *Florida Historical Quarterly* 70.2: 177–191.

Stokroos, M., 1992, 'In't nokkend wee der Ziele welt een Iraan'. Oude Graftekens op de Nieuwe Oosterbegraaf plaats. Bureau Monumentenzorg, Amsterdam, Netherlands.

Stone, G., 1991, Material Evidence of Ideological and Ethnic Choice in Long Island Gravestones, 1670–1800. *Material Culture* 23.3: 1–29.

Stott, A., 2003, The Woodmen of the World Monument Program. *Markers* 20: 1–29.

Strangstad, L., 1988, *A Graveyard Preservation Primer*. American Association for State and Local History, Nashville, IN.

Taafe, S., 1999, Commemorating the Fallen: Public memorials of the Irish dead of the Great War. *Archaeology Ireland* 13.3, 18–22.

Tait, C., 2002, *Death, Burial and Commemoration in Ireland, 1550–1650*. Palgrave Macmillan, London.

Tarlow, S., 1997, An Archaeology of Remembering: Death, Bereavement and the First World War. *Cambridge Archaeological Journal* 7.1: 105–121.

—1998, Romancing the stones: the graveyard boom of the later 18th century. In *Grave concerns: death and burial in England 1700 to 1850*, edited by M. Cox, pp. 33–43. Council for British Archaeology Research Report 113, York.

—1999a, Wormie Clay and Blessed Sleep. Death and disgust in later historic Britain. In *The familiar past? Archaeologies of Britain 1550–1950*, edited by S. Tarlow and S.West, pp. 183–198. Routledge, London.

—1999b, *Bereavement and Commemoration. An Archaeology of Mortality*. Blackwell, Oxford.

Tashjian, D., 1978, Puritan attitudes toward Iconoclasm. In *Puritan Gravestone Art II*, edited by P. Benes, pp. 37–45. The Dublin Seminar for New England Folklife, 3, 1978. Boston University Press, Boston.

Tashjian, A., and Tashjian, D., 1989, The Afro-American Section of Newport, Rhode Island's Common Burying Ground. In *Cemeteries and Gravemarkers. Voices of American Culture*, edited by R.E. Meyer, pp. 163–196. UMI Research, Ann Arbor, Mich.

Tashjian, D., and Tashjian, A., 1974, *Memorials for Children of Change: The Art of Early New England Stonecarving*. Wesleyan University Press, Middletown.

Taylor, L., 1983, *Mourning Dress: A Costume and Social History*, G. Allen and Unwing, London.

Thies, R.M., 2001, Quantrill's Three Graves and Other Reminders of the Lawrence Massacre. *Markers* 18: 1–29.

Thompson, J., 1998, Bodies, minds, and human remains. In *Grave concerns: death and burial in England 1700 to 1850*, edited by M. Cox, pp. 197–201. Council for British Archaeology Research Report 113, York.

Thompson, S., 1989, *Florida's Historic Cemeteries. A Preservation Handbook*. Historic Tallahassee Preservation Board, Tallahassee.

—1993, These Works of Mortuary Masonry: The Aboveground Tombs of St Michael Cemetery, Pensacola, Florida. *The Southern Quarterly* 31.2: 50–73.

—1997, *Historic African American and African Caribbean Cemeteries: A Selected Bibliography*. The Center for Historic Cemeteries Preservation, Tallahassee, FL.

Thompson, S., Brackner, J., and Lemmon, A.E., 1993, Historic Cemeteries in the Southern United States: A Preliminary Bibliography. *The Southern Quarterly* 31.2: 133–146.

Thursby, J.S., 2003, Mormon Temple Reproductions on Cemetery Markers. *Markers* 20: 312–332.

Trask, D., 1978a, *Life how short Eternity how long: Gravestone Carving and Carvers in Nova Scotia*. Nova Scotia Museum, Halifax.

—1978b, "J.W." Folk Carver of Hants County, Nova Scotia. In *Puritan Gravestone Art II*, edited by P. Benes, pp. 58–65. The Dublin Seminar for New England Folklife, 3. Boston University Press, Boston.

Trask, D., and McNabb, D., 1988, Research Report on the Graveyards of Kings County, Nova Scotia. *Markers* 5: 150–166.

Tucker, R.L., 1992, The Mullicken Family Gravestone Carvers of Bradford, Massachusetts, 1663–1768. *Markers* 9: 23–57.

—1993, The Lamson Family Gravestone Carvers of Charlestown and Malden, Massachusetts. *Markers* 10: 150–217.

—1995, The Joshua Hempstead Diary. *Markers* 12: 118–143.

—1997, The Pratt Family of Stonecutters. *Markers* 14: 134–157.

Turner, P.D., 1985, A particular mason: a study of Gloucestershire churchyards. *The Local Historian*: 16.7: 412–417.

Tyson, N., 1994, Candie Cemetery and its Monuments; Death and Self-Presentation in Nineteenth Century St. Peter Port, Guernsey. *Societé Guernesiaise Report and Transactions* 23.3: 599–626.

Ubelaker, D.H., 1989, *Human Sekeletal Remains: Excvavation Analysis, Interpretation*. Taraxacum, Washington D.C.

—1995, Historic Cemetery Analysis: Practical Considerations. In *Bodies of Evidence. Reconstructing History through Skeletal Analysis*, edited by A.L. Grauer, pp. 37–48. Wiley-Lis, New York.

Veit, R., 1995, 'A Piece of Granite That's Been Made in Two Weeks': Terra-Cotta Gravemarkers from New Jersey and New York, 1875–1930. *Markers* 12: 1–29.

—2000, John Solomon Teetzel and the Anglo-German Gravestone Carving Tradition of 18th Century Northwestern New Jersey. *Markers* 17: 124–161.

Vlach, J.M., 1978, *The Afro-American Tradition in Decorative Arts*. Cleveland Museum of Arts, Cleveland OH.

—1991, From Gravestone to Miracle: Traditional Perspective and the Work of William Edmondson. In *By the Work of Their Hands: Studies in Afro-American Folklife*, edited by J.M. Vlach, pp. 109–126. UMI Research Press, Ann Arbor, MI.

Wagner, A., 1967, *The Heralds of England: A History of the Office and College of Arms*. H.M.S.O. London

Walter, T., 1999, editor, *The Mourning for Diana*. Berg, Oxford.

Warren, N.H., 1987, New Mexico Village Camposantos. *Markers* 4: 115–129.

Watters, D.H., 1978, A Priest to the Temple. In *Puritan Gravestone Art II*, edited by P. Benes, pp. 25–36. The Dublin Seminar for New England Folklife, 3. Boston University Press, Boston.

—1981, *Eschatological Themes in Puritan Literature and Gravestone Art*. UMI Research Press, Ann Arbor, MI.

Watters, D.R., 1994, Mortuary Patterns at the Harney Site Slave Cemetery, Montserrat, in Caribbean Perspective. *Historical Archaeology* 28.3: 56–73.

Watts, L., and Rahtz, P.A., 1985, *Mary-le-Port, Bristol, Excavations 1962–63*. City of Bristol Museum and Art Gallery, Bristol.

Welch, R.F., 1987, The New York and New Jersey Gravestone Carving Tradition. *Markers* 4: 1–54.

West, E.S., 1989, The John Dwight Workshop in Shirley, Massachusetts, 1770–1816. *Markers* 6: 1–31.

Weston, D.A., 1989, editor, *The Sleeping City. The Story of Rockwood Necropolis*. Society of Australian Genealogists and Hale and Iremonger, Sydney, NSW.

Whinney, M., 1988, *Sculpture in Britain 1530 to 1830*. 2nd edition. Penguin, London.

White, B., 1998, The excavation and study of human remains: a view from the floor. In *Grave concerns: death and burial in England 1700 to 1850*, edited by M. Cox, pp. 247–251. Council for British Archaeology Research Report 113, York.

Whittemore, P., 2001, Waller Fecit, London. John Green, L.A.B. and William Augustus Waller. *Church Monuments* 16: 79–125.

Willats, R.M., 1987, Iron graveslabs: a sideline of the early iron industry. *Sussex Archaeological Collections* 125: 99–113.

Williams, G. Jr., 1992, The Center Church Crypt of New Haven Connecticut: Photographic Essay. *Markers* 9: 79–103.

—1994, Solomon Brewer: A Connecticut Valley Yankee in Westchester County. *Markers* 11: 52–81.

—2000, By Their Characters Shall You Know Them: Using Styles of Lettering to Identify Gravestone Carvers. *Markers* 17: 162–205.

Williams, M.M., and Williams, G. Jr., 1988, "Md. by Thos. Gold": The Gravestones of a New Haven Carver. *Markers* 5: 1–59.

Williams, R., 2000, Vanbrugh's India and his Mausolea for England. In *Sir John Vanbrugh and Landscape Architecture in Baroque England 1690–1730*, edited by C. Ridgway & R. Williams, pp. 114–130. Sutton, Stroud.

Willsher, B., 1985a, *How to Record Scottish graveyards*. Council for British Archaeology Scotland, Edinburgh, Scotland.

—1985b, *Understanding Scottish Graveyards. An interpretative approach*. Council for British Archaeology Scotland, Edinburgh, Scotland.

—1992a, Adam and Eve scenes on kirkyard monuments in the Scottish lowlands, *Proceedings of the Society of Antiquaries of Scotland* 124: 413–451. Almost exactly reproduced in 1995 as Adam and Eve Scenes on Kirkyards in the Scottish Lowlands: An Introduction and Gazetteer. *Markers* 12: 30–91.

—1992b, The Green Man as an Emblem on Scottish Tombstones. *Markers* 9: 58–77.

—1993, Scottish churchyard memorials of the eighteenth century. *The Local Historian* 23.2: 66–84.

—1996, *Scottish Epitaphs: Epitaphs and Images from Scottish Graveyards*. Cannongate Books, Edinburgh, Scotland.

Willsher, B., and Hunter, D., 1978, *Stones: eighteenth century Scottish gravestones*. Cannongate Books,Edinburgh, Scotland.

Wilson, J.S., 1992, Purchase Delay, Pricing Factors, and Attribution Elements in Gravestones from the Shop of Ithamar Spauldin. *Markers* 9: 104–131.

Wilson, S., 1998, *Garden of Memories. A Guide to Historic Forest Hills*. Forest Hills Educational Trust, Boston MA.

Winter, J.M., 1995, *Sites of Memory, Sites of Mourning: The Great War in European Cultural History*. Cambridge University Press, Cambridge.

Wright, J., 1980, Wisconsin's Wrought Iron Markers. *Markers* 1: 76–79.

Wright, R.A., 1988, Poems in Stone: The Tombs of Louis Henri Sullivan. *Markers* 5: 168–208.

Wrightson, K.K., 1998, Aboriginal Australian Burials in Christian Missions. *Markers* 15: 234–285.

Wrigley, E.A., and Schofield, R.S., 1981, *The Population History of England 1541–1871: A Reconstruction*. Arnold, London.

Wurst, L., 1991, "Employees Must Be of Moral and Temperate Habits": Rural and Urban Elite Ideologies. In *The Archaeology of Inequality*, edited by R.H. McGuire and R. Paynter, pp. 125–149. Blackwell, Oxford.

Yalland, Z., 1985, *Kacheri Cemetery, Kanpur. A complete list of inscriptions with notes on those buried there*. British Association for Cemeteries in South Asia, London.

Yates, T., Butlin, R., and Houston, J., 1998, *Carved Stone Decay in Scotland. Assessment Methodology Handbook*. Technical, Conservation, Research and Education Division. Historic Scotland, Edinburgh.

Yeoh, B.S.A., 1991, The Control of 'Sacred' Space: Conflicts Over the Chinese Burial Grounds in Colonial Singapore, 1880–1930. *Journal of Southeast Asian Studies* 22.2: 282–311.

Yeoh, B.S.A., and Hui, T.B., 1995, The Politics of Space: Changing Discourses on Chinese Burial Grounds in Post-War Singapore. *Journal of Historical Geography* 21: 184–201.

Young, S.E.J., 1998, Archaeology and smallpox. In *Grave concerns: death and burial in England 1700 to 1850*, edited by M. Cox, pp. 190–196. Council for British Archaeology Research Report 113, York.

Zaniello, T.A., 1980, The Keystone of Neoclassicism: Freemasonry and Gravestone Iconography. *Journal of American Culture* 3.4: 581–594.

Zelinka, S., 1991, *Tender Sympathies: A social history of Botany Cemetery and the Eastern Suburbs Crematorium*. Hale and Iremonger, Sydney NSW.

Zelinsky, W., 1976, Unearthly Delights: Cemetery Names and the Map of the Changing American Afterworld. In *Geographies of the Mind: Essays in Historical Geography*, edited by D. Lowenthal and M.J. Bowden, pp. 171–195. Oxford University Press, New York.

—1990, A Toponymic Approach to the Geography of American Cemeteries. *Names: Journal of the American Name Society* 38.3: 209–229.

—1994, Gathering Places for America's Dead: How Many, Where, and Why? *The Professional Geographer* 46.1: 29–38.

Zucchi, A., 1997, Tombs and Testaments: Mortuary Practices During the Seventeenth to Nineteenth Centuries in the Spanish-Venezuelan Catholic Tradition. *Historical Archaeology* 31.2: 31–41.

Index